W9-BBD-906

# SINKABLE

Also by Daniel Stone

*The Food Explorer*

# SINKABLE

## OBSESSION, THE DEEP SEA, AND THE SHIPWRECK OF THE *TITANIC*

DANIEL STONE

DUTTON

DUTTON
An imprint of Penguin Random House LLC
penguinrandomhouse.com



Illustrations by Matthew Twombly

LIBRARY OF CONGRESS CATALOGING-IN-PUBLICATION DATA
has been applied for.

ISBN 9780593329375 (hardcover)
ISBN 9780593329399 (ebook)

Printed in the United States of America
1st Printing

BOOK DESIGN BY LAURA K. CORLESS

*For my grandparents, and theirs, too,*
*who took ships over seas so that*
*I can fly across them*

# Contents

*Marine salvage: A science of vague assumptions based on debatable figures taken from inconclusive experiments and performed with instruments of problematic accuracy by persons of doubtful reliability and questionable mentality.*

—CAPTAIN CHARLES A. BARTHOLOMEW, U.S. NAVY

# Author's Note

If I asked you to guess, a shot in the dark, how many shipwrecks sit at the bottom of the oceans as relics of accidents past, what would you say? I have some advantage in asking this question because for several years I have asked almost everyone I've met. All of human history, all of earth's water, how many shipwrecks?

Guesses range widely from very tiny to very big. But what they all have in common is that they are all too small. They are most underestimated by kids, who can't be blamed; their education of early human sea travel begins with the relatively recent eras of Magellan, the Spanish Armada, the *Nina* and the *Pinta* and the still-missing *Santa Maria*. Maybe they've heard of the *Titanic*, but likely through the lens of Leonardo DiCaprio or, more likely, Celine Dion. Adults are wrong too, but they're wrong by less. Most adults think of ships in the modern sense, such as the USS *Arizona* they saw in Pearl Harbor while on vacation or the *Costa Concordia* from the news a few years ago. Looking backward, our historic understanding of ships clings to a few historical nuclei: Captain Cook, Columbus, the *Maine*, Ernest Shackleton, the *Lusitania*, or, again, the *Titanic.*

"Maybe a quarter million" was the highest estimate I ever heard, from a friend who is a U.S. marine and the kind of guy who spends a lot of time in naval museums. It wasn't a bad guess, but when you account for all of human history and our planet's mostly aquiferous surface, the number is substantially higher. According to a UNESCO estimate, the number of ships that sit in underwater graves, wasting away year after year, is an incredible three million.

When the *Titanic* sank in 1912, it plunged an astounding two and a half miles and hit the seafloor at more than thirty miles per hour. Its ocean grave was so remote that its location remained a mystery until 1985, when a team that had the benefit of government-developed submarines and deepwater crafts was able to take some blurry snapshots. It took seventy-three years, almost an entire human life-span, to find the most illustrious and fascinating shipwreck of all time.

This is the story of what happened *after* the night the ship sank—how the *Titanic* changed the world and how the world longed desperately to piece it back together.

Shipwrecks have a habit of attracting colorful and unusual characters. Those characters tend to be overwhelmingly male and most often white—a reflection of a historically macho industry that runs on expensive machinery and huge sums of money. Many will tell you that they work in the most punishing environment on earth (true) and also that working in the deep sea is harder than working on Mars (possibly true). Yet just as many are armchair enthusiasts, amateurs, and white belts, whose sole credential is their obsession. They get drawn into races, fights, and elaborate lawsuits. The odd quirk that ships are typically female is a holdover from times when vessels and their wrecks could—and still can—stand in as full-blooded companions.

In the process of writing, I interviewed dozens of historians, salvage professionals, wreck experts, and lawyers, people who earn their livelihoods drilling for oil, hunting for treasure, investigating accidents, and filming documentaries. There's a common characteristic I noticed they share, perhaps best described as a polite but profound impatience for your crap. They have work to do, and they've seen more otherworldly stuff than you ever will. They're usually under pressure from rich investors and impatient scientists, and they carry an enormous amount of risk managing equipment, timelines, and people's lives.

But every single one will eventually soften up and readily dish about the granddaddy of them all. In cultural lore, the *Titanic* is the wreck around which all others orbit. The same way a pop musician can't escape the influence of the Beatles or Michael Jackson, shipwreckers can't bypass the brightest star. In scientific terms, the *Titanic* embodies the waves of technological growth, failure, and advancement during its life-span above water, and relentless obsession and elaborate deep-sea engineering below.

This is not, in the words of Walter Lord, who wrote *A Night to Remember*, the seminal volume on the *Titanic*'s final night, "another book about the *Titanic*." This is a look at our oceans and the junk we've left in them. It is a yarn about the oddballs and misfits who devote their lives to wayward ships. And it is a deep dive into the waters of our planet and what lurks, in every sense, just below the surface.

DANIEL STONE
*Santa Barbara, California*
*April 2021*

# SINKABLE

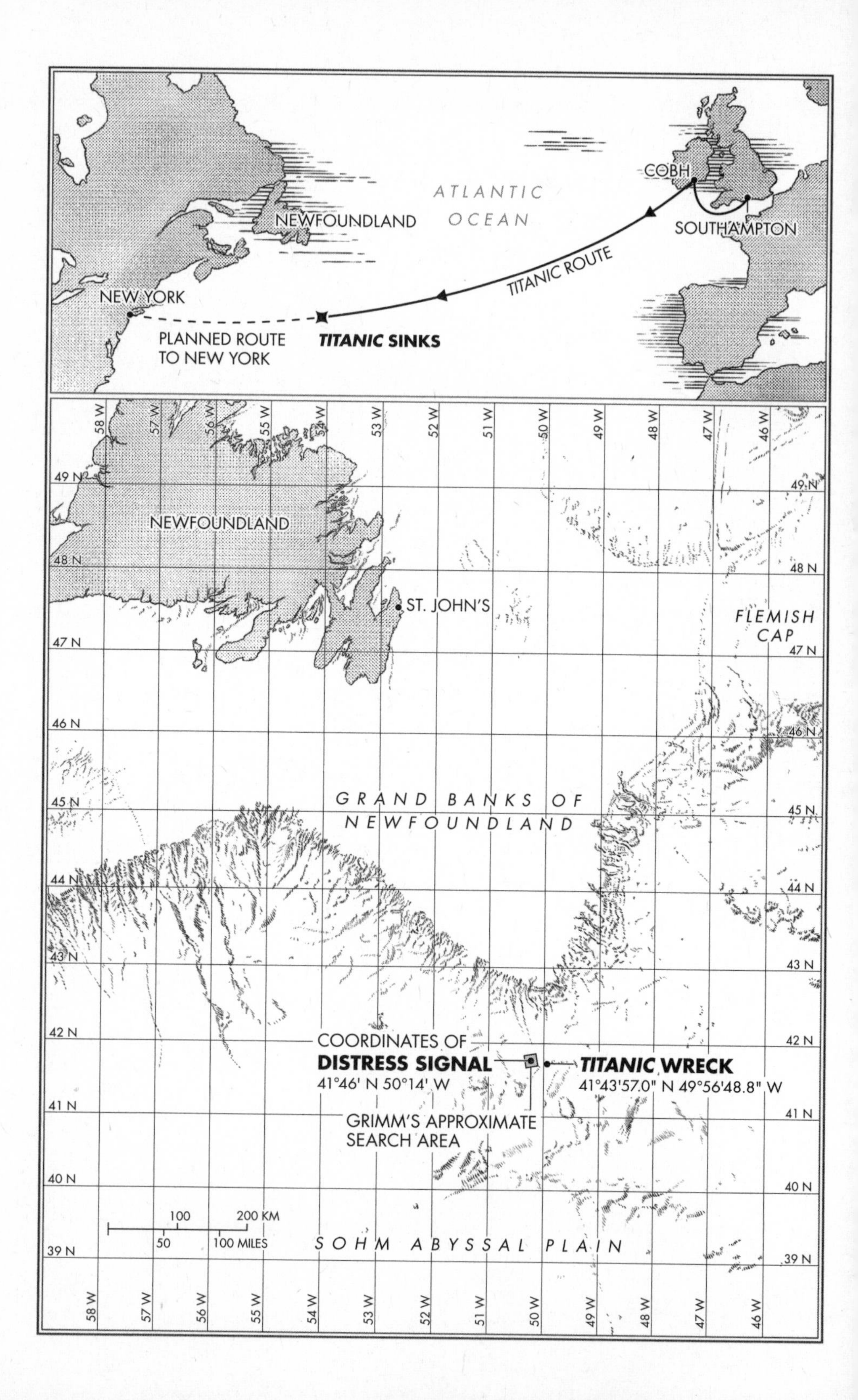
ATLANTIC OCEAN
NEWFOUNDLAND
NEW YORK
COBH
SOUTHAMPTON
TITANIC ROUTE
PLANNED ROUTE TO NEW YORK
*TITANIC* SINKS
NEWFOUNDLAND
ST. JOHN'S
FLEMISH CAP
GRAND BANKS OF NEWFOUNDLAND
COORDINATES OF
**DISTRESS SIGNAL**
41°46' N 50°14' W
*TITANIC* WRECK
41°43'57.0" N 49°56'48.8" W
GRIMM'S APPROXIMATE SEARCH AREA
SOHM ABYSSAL PLAIN
100
200 KM
50
100 MILES
58 W
57 W
56 W
55 W
54 W
53 W
52 W
51 W
50 W
49 W
48 W
47 W
46 W
49 N
48 N
47 N
46 N
45 N
44 N
43 N
42 N
41 N
40 N
39 N

# Prologue

When snow falls, the properties of water perform a delicate dance. Snowflakes fall like dominoes fall. A piece of dust forms a crystal, and the appearance of that crystal attracts more crystals until they form long dendrites around the speck of dust like ants around a piece of chocolate. As long as the growing snowflake remains lighter than air, it will float. But as soon as one extra crystal crosses the tipping point, the structure will succumb to gravity and fall.

Snow tends to fall in places where other snow has already fallen. And even though every snowflake is different, they're not as unique as we've been told. They start as spheres and form tendrils to diffuse heat. Cold temperatures produce flakes that look like bullets or needles. Extra-cold weather is when you find the classic shape of a six-sided prism, or the fernlike crystal with six radiating branches. For some reason, there are always six.

It was probably this form of fernlike snow that fell one day, fifteen thousand years ago, on the frozen ice sheets of Greenland. The landmass was already covered in ice two miles thick. With time, the fresh flakes descended into the ice, hidden from daylight and compressed by pressure to a third of their original size.

Fitting with geology, thousands of years passed and nothing happened. Snow that started as flakes was transformed to dense glacial ice as it moved quickly, about four miles per year, toward the west coast of Greenland. Ice weakens as it nears the coast, because every day, particularly in the summer, enormous walls of ice flake off the glacier and fall into the ocean.

This is how ocean icebergs form. But it was one particular iceberg that fell in the summer of 1909 that would drift toward infamy. Around too briefly to have a name, this iceberg was more than two miles wide and one hundred feet tall at its birth, big enough to dwarf the Colosseum in Rome and all the pyramids put together, at least before it started melting. It would tower over the largest steamship ever conceived, which was also formed in that summer of 1909.

At the time, humans knew little about the behavior of icebergs, except that most melted somewhere in the Arctic Circle. John Thomas Towson, a scientist devoted to ship navigation who wrote a book called *Practical Information on the Deviation of the Compass*, observed in 1857 that icebergs were no different—and no softer—than rocks formed over millennia by time and pressure. Towson knew that icebergs posed an existential danger to the wooden hulls of nineteenth-century ships. Steel hulls were invincible, he said, but that was based on assumption, not experience. Such an extreme number of icebergs traveled south through the east strait of the Grand Banks in eastern Newfoundland that in 1912 the U.S. Coast Guard nicknamed the area "iceberg alley."

For three years the icy mass bobbed and weaved in Arctic waters. At one point, it traveled north and spent the summer of 1910 farther toward the north pole. Then it caught the Labrador current, which carries freezing water south. Most icebergs melt within their first year. A few last two. Only a handful last three because, eventually, the Labrador current meets the warm waters of the Gulf Stream, which acts as an oceanic microwave. Only 1 percent of northern hemisphere icebergs survive this desert zone, and finally, only one in several thousand would make it to 41 degrees north, the same latitude as New York City and directly in the path of transatlantic ships.

Icebergs had struck ships as long as there had been ships to strike, but few icebergs had been as lucky as the one that felled the largest passenger liner ever built, particularly because, on April 14, 1912, that iceberg was nearly gone. After three years adrift, the icy mass likely had one week to live, two at most. As icebergs melt from the bottom, they grow top-heavy and flip, followed by more erosion and more flipping, until eventually, when they've been reduced to the size of a basketball, they're constantly flipping until there's nothing left.

Any other week and a ship nobody believed could sink would complete its maiden voyage and turn around for its ho-hum second one. Any other day and the iceberg would've been a mere fraction of its dangerous size. Any other hour and it would've been hundreds of feet away. But the ship waited for nothing, and the ice knew nothing to wait for, and the ingenuity of humans at the dawn of modern invention succumbed, rather incredibly, to the force of several crushed-up snowflakes as hard as rock.

Chapter 1

# SHIPFALL

When Ok-Khun Chamnan, a diplomat from Siam on his way to Portugal, saw seawater filling the hull of the ship he was traveling on, he knew he and his fellow passengers were done for. In April of 1686, the ship, its name lost to history, sailed too close to the rocky shallows of Cape Agulhas off southern Africa. Ocean waves lifted the ship and slammed it on the rocks. The hull cracked on all sides as it was raised up and again plunked down hard. Chamnan watched the crew cut down the masts and throw the guns overboard, the resignation of a lost cause. But it was too late.

"The water [was] entering in abundance," recalled another survivor. Water filled the first deck, followed by the gunner's room, to the captain's cabin, and finally to the upper decks. "Our ship at last sunk quite down into the Sea," the survivor wrote. "It would be a hard task to represent the astonishment, terror and consternation that seiz'd upon every Heart in the Ship. Nothing now was heard but cries, sighs and groans."

Many passengers aboard the ship died. But several lived, including Chamnan, who for the rest of his life told an embellished tale of the experience to every willing audience. Crawling over rocks and fierce seas, he would say, the survivors made it to shore, where, wet and naked, they found nothing but more rocks and rain. Wild animals nipped at their heels. They wandered for over a month, eating lizards, running from lions, and drinking from puddles. Eventually, they made it to the Dutch trading station at the Cape of Good Hope and were rescued.

This tale, one of the earliest first-person accounts of a shipwreck, was preserved for centuries because it was written down. But there's little about this story that makes it unique. To be battered and beaten at sea on a sinking ship is a condition not special to any era. Boarding an oceangoing vessel in the seventeenth century brought the risk of danger and death, the same as it did in the century before and every one to follow. What all ships have in common, from a three-hundred-year-old merchant ship to the most modern aircraft carrier, is that, eventually, they fail.

Flooding is the most common reason ships sink. Ships float because they're lighter than the weight of the water they displace. But violent waves and a flooded deck can shift the balance, even slightly, and make a ship that was once lighter than water suddenly heavier. Every year, as many as thirty large ships go missing at the hands of large waves, some as tall as sixty feet, to say little of the uncounted sailboats, yachts, and leisure pontoons that sink every day. Nearly all escape even a passing mention in the news. "Imagine the headlines if even a single 747 slipped off the map with all its passengers and was never heard from again," writes Susan Casey, a chronicler of the world's largest waves, which, to this day, still swallow the most advanced steel vessels.

After flooding, sinkings are the frequent result of ground strikes, or, less often, collisions with other ships. For a long time, this was intentional. A ship's design—oblong with pointed ends—was for it not only to swiftly cut through the water but also to ram other ships at their weak center. Before cannons, guns, or even catapults that were reliably accurate, naval battles were decided by the strategy of who could more quickly position their ship in an offensive position and row like hell.

For every ship that hits an iceberg or strikes another vessel, there are thousands more that run into rocks or get moored irreversibly in mud. Some reef systems are especially punishing, like the Seven Stones Reef off the west coast of England, or the Kenn Reefs east of Australia, or the rocky straits of Lombok and Makassar in Southeast Asia. Each has claimed thousands of ships, and because they sit in shallow reef systems, they're especially popular among wreck divers.

The damage can be mutual. Trying to measure how many ships scrape the ocean bottom is like asking how many cars tap bumpers while parallel parking. Unless the damage is severe, the only witnesses are fish and whales, who must have their own feelings about ship strikes. Usually it's not the well-known reefs that are the most dangerous but the rocky outcrops in unassuming waters that prey on unsuspecting ships. In the span of eight days in August of 2010, a cargo vessel and two container ships all ran aground in the coastal waters of India, causing two hundred containers to fall into the sea and creating an oil spill visible for miles. Even a ship with a delicate name like *Belle Rose* can be ruthless. In 2016, an error blamed on the crew caused the destruction of seven acres of coral reef off

Malapascua Island in the central Philippines, the world's top habitat for thresher sharks.

Then there are the wrecks caused by imbalance, a dull demise but still deeply frightening because of its suddenness. All floating objects have what's known as a metacenter, which can be pictured as a vertical line drawn upward through the center of the ship. The metacenter indicates a ship's center of gravity, which shifts with every wave. Container ships have to factor imbalance into how they're loaded and how they move. Stacking containers too high increases a ship's side profile, a measure known as its windage, which can be like driving a semitruck through a windy canyon. Pushed too far by a monstrous swell or a gust, the ship topples over. Accidents of imbalance can be embarrassing for captains because they're often caused by poor loading or shoddy engineering. It took Sweden more than three hundred years to laugh about its most famous wreck, the *Vasa*. The ship was so asymmetrically designed that a gust of wind during its maiden voyage in 1628 caused a list to one side, which filled the lower gunports with water, which was all it took to sink the *Vasa*.

One of the most bizarre phenomena is when an ordinary-looking ship sinks for no reason. This is sometimes the result of liquefaction, a process that occurs when solid cargo turns to liquid due to the vibration of the engine. You might imagine carrying a bucket of mud that jiggles as water rises up, and how you'd be knocked off-balance by the sloshing. Landlubber truckers are familiar with this principle. Carrying solids is easy, but if they break suddenly while moving a dozen tons of oil or glue, it'll slosh forward and yank the truck back. It's worse for ships, which get pulled in all directions. In May of 2005, the *Hui Long*, a midsize cargo vessel in benign conditions off the coast of Sumatra, was carrying fine-grained minerals and began to list without warning as the cargo began to shift. Within thirty minutes, the list was so steep the captain gave orders to abandon ship.

People are the dominant reason ships sink. The weird world of shipwrecks is filled with tales of overzealous captains, unrealistic schedules, hubris in the face of dangerous weather, and weary crews. One bad decision begets another, and eventually the lower decks are taking on water. That's usually the beginning of the end, as it was on April 15, 1912. One shipwreck among millions, plucked from a slow recession into obscurity and instead transformed into a cultural symbol that became, through the lens of time, a turning point in history.

° ° °

People who study shipwrecks for a living are often tired of talking about the *Titanic.* It was interesting, they'll grant, and some famous people died. But there's little about the fate of the most domineering ship of twentieth-century folklore to warrant its disproportional place in the cultural zeitgeist.

Large ships had failed before, many from collisions with icebergs. In 1854, the SS *City of Glasgow* disappeared on its way from Liverpool to Philadelphia, along with four hundred eighty people. The SS *Naronic*, en route from Liverpool to New York in 1893, also vanished, with seventy-four aboard. Not only was the *Naronic*'s fate met with apathy, it was also a complete mystery until messages were later found floating in bottles, apparently written by passengers who blamed their disappearance on an iceberg strike. Icebergs were such a common scourge of the North Atlantic that by 1912, most experts were relieved that collisions with icebergs appeared to have declined. Prior decades saw as many as seven strikes each month; by 1910, there were only about four per year.

A high death toll wasn't it, either. Other wrecks had drawn greater losses of life, like the Chinese junk ship *Tek Sing*, whose sixteen hundred passengers were killed in 1822 when it ran aground in the South China Sea, or the French munitions ship *Mont-Blanc*, which sank in 1917 after an explosion so fierce in the harbor of Halifax, Nova Scotia, that falling debris killed more than two thousand people *on shore*.

When it comes to explaining the *Titanic*'s enlarged relevance, there are the nebulous explanations about human confidence, about a symbol of a new era and the embodiment of modernism, a boat against the current borne back ceaselessly into the past. The satirical newspaper *The Onion* put a fine point on it in a retro edition headline, "World's Largest Metaphor Hits Iceberg." These theories carry water, but they also too easily dismiss the fact that the *Titanic* didn't become an instant metaphor or a cultural realignment in its day. It was a tragedy, one of many in an uncertain era, that happened to kill mostly rich people.

The *Titanic*'s quick growth into a news story big enough that it warranted *The New York Times* renting out an entire floor of a hotel to cover the sinking was based on one particular and often overlooked fact. It wasn't that fifteen hundred people died, but that seven hundred people lived. Had every last soul been dragged to the bottom of the Atlantic, it would've joined the voluminous annals of devastating maritime tragedies. Memorials would've been held, insurance checks would have been paid, and the world would've moved on. But a tragedy with hundreds of survivors meant there would be hundreds of gripping accounts of the ship's final moments, the wrestling and jockeying, the rescued and the abandoned, the brave and the weak. There were many—and at times conflicting—tales of valor, cowardice, fear, triumph, and horror for the public to adjudicate. History, after all, isn't told by the dead.

What's more, on account of women and children being granted the limited spots to escape, many of the survivors were young, and their youth ensured decades of tellings and retellings of their stories. Eva Hart was seven in 1912 when she stepped off the *Titanic* into a lifeboat with her mother. She realized years later that the barely three-day experience during her childhood would be the seminal moment of her life. Like many survivors, she struggled to shed her association with the disaster, which had killed her father, as the centerpiece of her identity, and when she realized no amount of changing the subject or politely declining to answer the same questions again and again would be sufficient, she embraced the role. She spoke out against the "ridiculous" shortage of lifeboats and, decades later, about the "greed" of the vultures who wanted to salvage the wreck site. Throughout her life she monetized her tragedy in speeches and a book and transformed into an Oprah-like figure who turned her early-life trauma into a message of resilience, perseverance, and hope.

Other survivors dwelled in primal human emotions, even among people who had already heard the story ad nauseam. "The agony of that night can never be told," Charlotte Collyer, a thirty-year-old wife and mother, would write in a letter to her mother after she survived. And yet, she would also tell people, "I shall never forget the terrible beauty" of the *Titanic* in its final moments as she watched its famous twenty-three-degree tilt and its ferocious snap. The searing memories of such horror were too complex for a person to process in a single lifetime, and this mix of confusion, pain, and awe was like a flame. No one could look away. Who said what, who argued with whom, and all the while, what the band was playing. The details have been turned over and over, and for some reason, even when you know how the story ends, it never gets old. (This may also explain the phenomenon, unique in 1997, when moviegoers went to see James Cameron's film multiple times in theaters, never able to get enough.)

The most compelling explanation for the *Titanic*'s outsize cultural staying power is the simplest. And in the case of a century-old shipwreck among thousands of other deadly boating accidents, the rationale seems to come down to something timeless: good storytelling.

Isolate all the components that the *Titanic* shared with other ships and other disasters—iceberg strike, loss of life, human overconfidence—and what's left are the same components that make any story in any era worth hearing: high stakes, an intricate but linear narrative arc, emotional turns of tragedy and triumph, and a dollop of suspense, even still, about what exactly happened. Taken together, it's little wonder why anyone who touches the *Titanic* risks getting caught up in its endless current. Like barnacles on a hull, some people just want to be near it.

ooo

How can you be sure about the way a ship sank? You can study the accounts of witnesses or simulate the conditions of a ship in a storm. Many passenger ships now have voyage data recorders, the equivalent to the black boxes in airplanes, which record a vessel's final gasping hours. But get past the *what* that caused a vessel to sink and it becomes a marvel to study *how* ships sink. How they fall through the water, the twists and pirouettes, the grace followed usually by a crash.

Every ship can sail thousands of times and carry millions of people. But when it sinks, it sinks only once. There's worldwide certainty about what caused the *Titanic* to sink. But then what? Did it twist and then turn, or turn first? How did it land on the seafloor, and at what speed? It's reasonable to wonder if it matters. It sank, people died, it's gone. But studying what shipbuilders refer to as "shipfall" informs how future ships might be better built and how to fortify them from the sort of destruction that struck the most famous one.

There are many theories about *how* the *Titanic* sank, how it fell through the various ocean layers known as the water column and crashed violently into the deep-ocean seabed. One of the most advanced theories, based on computer modeling and nautical forensics by *National Geographic* and a handful of scientists, holds that the *Titanic* began its fall slowly before picking up violent speed and pressure as it fell. The calculations are based on simple physics equations of mass, ocean current, and distance. Plug them into modeling software and the *Titanic* takes on a clumsy elegance. When the *Titanic* ripped in two, the bow swung down, held to the stern by a thin layer of steel the way two halves of a cut tree still hold fibers that are hard to break. That lasted mere seconds before the rupture was complete.

For barely a moment—and for the last time—the ship traveled through the thin layer of water, from zero to six hundred feet, known as the surface ocean. Had a passenger taken a deep breath, he might have survived this depth and the growing clamp of pressure. But no one survived what came next.

The bow, the pointed front of the ship, entered a free fall, its sharpest edge steering straight for the ocean bottom. Above ground, this would be known as planing, when a bird stiffens its wings and glides with no effort. Underwater, though, it's just falling, sinking, or, put scientifically, succumbing to the laws of density and gravity. Even in water, this happened fast. As a ship, the *Titanic* was designed for a maximum surface speed of twenty-one knots, or twenty-four miles per hour. As a newborn shipwreck, it amassed greater power, accelerating to a terminal thirty miles per hour, while nearly three hundred pounds per square inch of pressure crushed every possible air pocket in the couch cushions, the wine bottles, and the narrow space between a painting and its glass. The most that early-twentieth-century scientists knew about pressure at these depths had been learned by lowering a length of rope a mile deep. When it was pulled back to the boat, the end that had touched bottom was half as thick as when it started.

The *Titanic*'s stern, meanwhile, lagged behind. It received a forceful bob upward when the bow severed its hold, but this trajectory was brief. Barely a minute after the bow started downward, the stern began its own undersea journey with a vertical lean and a slow succumbing from the surface. It fell slower, dragged by the angle at which it fell, blunt end first. Bigger air pockets in the stern, occupied by first-class cabins, smoke rooms, and the grand staircase, gave way to an implosion so forceful it was supposedly heard by survivors floating hundreds of feet above.

Had it been daytime instead of night, sunlight would've illuminated everything to this point. The first thousand feet of water, known as the epipelagic zone, from the Greek words for *the top of the sea*, is home to almost all fish, kelp, reefs, and marine animals known to science, all of which benefit from the warmth and photosynthesis of the sun. Only one time has a human swum below this zone and lived to tell about it, an Egyptian army officer named Ahmed Gabr. In 2014, Gabr scuba dove a fifth of a mile, a journey that took him twelve minutes down and fifteen hours up to decompress. At his deepest, Gabr withstood more than four hundred forty pounds on every square inch of his body, making him feel extremely heavy and cold. That was the worst part. The best part, he said, was on the way up, when a baby oceanic whitetip shark hung out with him for six hours.

Even on the brightest day, water below six hundred feet turns black. And from here, both halves of the ship fell with accelerating speed as all the handrails, the lampposts, the mast, and any fastened debris were pulled violently off. Below six hundred feet, the ship entered the twilight zone, a three-thousand-foot stretch of the water column sparsely populated with weak gelatinous fish that tend to eat whatever detritus falls from above or is buoyant enough to float up from below. There would be fewer of them as the ship entered lower depths, known as the midnight zone, or as the Greeks called it, the bathypelagic, from the word for *deep*.

Despite its lack of hospitality for human life, the midnight zone is quite pretty. The fish, mollusks, crustaceans, and jellyfish that survive this deep find their way in the darkness by creating their own light using chemical bioluminescence. Their sprinklings of color, most of it blue, help attract prey. Swimming too far at these depths is energetically expensive, so some fish in the midnight zone evolved other qualities to make life easier, like red bodies that in the absence of red light give them a noir effect of invisibility to predators. Other fish called tubeshoulders release clouds of luminescent fluid to lure smaller organisms. Once they're close, they bite with backward-pointing teeth—an evolutionary quirk designed to minimize effort in a part of the ocean that's hard enough to begin with.

Into this otherworldly constellation rushed the broken bow of a ship falling pointed-side down and then stabilizing to be right-side up. Engineers have since chalked up the *Titanic*'s changing position to drag; the smooth underside of a ship will cut through the water faster than its textured top, not to mention air bubbles, desperate at this punishing depth to escape the wreck through the easiest path possible, which was the top.

The stern, meanwhile, followed the bow quickly. First it imploded, and then it fell into a spiral, a bit like a helicopter blade, as the air bubbles acted as propulsive jets that pushed the structure around and around. Eventually it stabilized, having ejected most of the remaining air. Then it fell right-side up.

The best possible position for a ship to fall through the water is flat, identical to its above-water position. This maximizes the surface area of the vessel when it strikes the bottom, which minimizes to every extent possible the explosion expected if a vessel strikes the floor like a missile. There are humanitarian reasons to optimize shipfall, as well as financial ones. A vessel largely intact underwater makes it easier to extract dead bodies, treasure, or valuable equipment, like a voyage data recorder. More than one hundred years after the *Titanic* sank, scientists still believed it was possible to excavate the ship's famous telegraph, which begged for help from nearby ships.

The way a ship falls is also crucial to knowing how it lies, and where it lies helps determine how to explore it. Anyone who clung to the ship or was trapped inside one of its crevices was dead before it reached the seabed. But if there was even one sentient cell left, its final trauma would have been the ground strike followed by three other powerful forces. The first was the ship "breaking its back," a technical term for when the long steel plates that made up the ship's keel and deck panels received a forceful concussion, the same way the steel in a crashing car buckles on impact.

A moment later, the ground strike caused the second force, an enormous burst of water pushed violently from inside the wreck outward. Researchers found in 2012 that this outburst was enough to blow off the ship's front hatch cover, a manhole-size piece of metal held down by more than a dozen large bolts. Even at twelve thousand feet, a hydraulic burst would feel to a human like sitting under Niagara Falls.

The final insult, after the iceberg, the implosions, the ground concussions, and the hydraulic outbursts, was a powerful jet stream of water that struck the ship from above. This phenomenon is known as the downblast effect and occurs when something sinking pulls the water behind it, filling in the momentarily empty space it leaves in its wake. The downblast effect helps explain the myth of suction, the mistaken notion that one must swim away from a sinking ship to avoid being pulled down with it. It's a notoriously hard thing to test, but when the suction effect has been observed, it's only when large wrecks have started sinking at high speed, putting in motion the water column behind them.

Again, why does this matter? You might think differently if you're ever on a sinking ship. There's little scientific consensus about what you should do, so generally, do whatever will save or prolong your life. But captains and lifeguards tend to agree that the best thing you can do if your boat is sinking is to stay on the endangered boat as long as possible. Climb to high ground if you can, and even higher ground after that. If you're within sight of land or another ship, take off your clothes for the same reason that no Olympic swimmer ever won a race wearing waterlogged blue jeans. And then, once you touch water, start swimming toward a lifeboat or a piece of debris. Try not to be directly above the ship after it's underwater, since it will probably be releasing air bubbles, which make the water less dense, and thus make it harder to swim. But if you are above the ship as it sinks underwater and all else fails, take the biggest breath you can, push off the boat, and kick up with everything you've got.

° ° °

Women and children were supposed to be rescued first. And for the most part with the *Titanic*, they were; 70 percent of women and children made it off the ship alive. But this was a historical anomaly, because for almost all other sinking ships, women and children go last, if they're rescued at all. In 2012, two Swedish researchers studied eighteen maritime disasters involving fifteen thousand people of more than thirty nationalities over three centuries and found that not only are men's survival rates twice as high as women's, but that children fare worst of all—just 15 percent tend to make it off alive. What's more, while the notion of a captain "going down with the ship" was true in the case of the *Titanic*, it's far from historical reality for almost all other shipwrecks. Captains and crew survive at significantly higher rates than passengers, a disparity likely explained by the fact that professional seafarers have more advanced survival skills and knowledge of a ship's layout, but also an indictment that they don't stay to help once it's every person for themselves. The myth of women and children first was a creation of the British elite, who used it for centuries as an argument against women's suffrage. Why do women need to vote, they claimed with straight faces, when even when facing death, men will put the interests of women first? The argument largely worked. Sixteen years and hundreds more shipwrecks occurred, ones where women's lives weren't prioritized or even assisted at all, before women in England were granted the same voting rights as men.

To survive the *Titanic* was better than the alternative that befell more than fifteen hundred passengers. But it was still no picnic. First there was the cold and the wind, and then the sight of horror unfolding in front of the survivors. The men rowing each lifeboat rowed harder and harder, as though trying to escape the looming fog of disaster while scanning the horizon for any sign of other boats. There were also the arguments: Should each lifeboat return to the site to pick up more people in the water? Lifeboat number one, commonly known as the captain's boat and with a capacity of forty, left the ship with only twelve. A fireman named Charles Hendrickson was one of the twelve and spent years after the disaster claiming that he was the only one on the lifeboat to propose going back to rescue more people still clamoring in the water. He was savagely overruled, he claimed, which brought shame upon the other eleven. The story was exaggerated and later refuted by an investigation by the British Board of Trade. But the damage had been done in the press, and one of the eleven, a fashion designer named Lucy Duff-Gordon, lived the rest of her life trying to salvage her and her husband's reputations as heartless cowards.

Help arrived on account of telegraph messages that grew increasingly frantic as the waterline rose. The first call was "CQD," which had evolved from the earlier radio call CQ to precede messages to everyone in range. But in the early years of the century, the British code CQD conflicted with the American distress code of NC and the German code of SOE. The Italians, meanwhile, used SSSDDD. None of these letters meant anything or were acronyms for longer phrases. And neither did the eventual international replacement, which had been decided at the urgent 1906 International Radio Telegraphic Conference that agreed a standard was needed. From then on, ships in distress would use SOS, chosen for its distinct mix of short and long bursts, and not to mean "save our ship," "save our souls," or any other convenient phrases retroactively attributed to the call.

The death toll would have been higher if not for the telegraph. The technology had its limits—just three hundred miles in daylight and a little more than double at night thanks to radiation changes in the atmosphere—but to anyone who had started their life in the previous century, this was no less exciting than teleporting one's own body to a faraway ship. Prior to this, the best way two ships could communicate was by colored flags or bursts of light, but even the clearest visual cues were quickly obscured by the curvature of the earth. The giddiness over the telegraph led it to be quickly trivialized. The way the system worked, with a telegraph blast from one ship received by all passing boats, was most useful for matters of safety and navigation. But for the rest of the time, first-class passengers were permitted to use the system to send greetings to all passing boats, and they did so with messages as simple as "Hello!" and "Good day!" No one had yet figured out a separation of channels that might divide superficial small talk from operational guidance.

In the days after the disaster, more limits were put on the telegraph based on how it performed in a crisis. The breaking-news announcement that would wash over the entire world started as a ship-to-ship game of telephone that was predictably distorted by the time word arrived in New York. The *Carpathia*, traveling from New York to modern-day Croatia before rerouting to the wreck site, had sent the first reports of the *Titanic* in danger before its officers knew just how devastating its predicament was. The message was picked up by radio operators on ships as far as five hundred miles away, who had no idea in the early hours of April 15 whether the *Titanic* sustained minor damage and was being towed to port, as *Carpathia* had initially reported, or if something much worse had happened.

Meanwhile, the stunted telegraph messages began to build a wave that would crest over the entire world. In the Associated Press offices in New York, a city editor named Charles Crane sat with his feet on his desk reading a novel by H. G. Wells when a colleague bolted into the newsroom waving a wire message that read, "Reported *Titanic* struck iceberg." That was correct, but almost everything that came next wasn't. Almost all newspapers reported an erroneous account of the event, from word that the ship had survived and was on its way to Halifax to reports that everyone on board was presumed dead.

Like any breaking-news story, the coverage followed a well-worn formula. First, there was the pinning down of details of what happened, who was affected, and when. Then there were the human-interest stories about the lives lost and the final moments of those swallowed by the sea. And then, predictably, the finger-pointing.

No one—not the reporters covering the story, not the White Star Line executives who would soon be answering for their mistakes, and not the survivors on the *Carpathia* or the crowds of rubberneckers and looky-loos who met them in New York—could imagine the oversize, century-long legacy awaiting the *Titanic*. The coal workers, department store managers, and housewives who read about all of it in the newspapers never realized that the events of that moment and their place in the story would reverberate for decades.

Lost in the fervor of the human element of the story, however, was the central character in the whole tragedy. In those early days when the ship was settling into its underwater tomb, the mangled mess of steel and wood creaked with the current and shuffled until it became snagged on something and was sucked into the mud. Resigned to its fate, the *Titanic* released the final bubbles of air still locked inside. And when the last one rose up and popped on the surface, the ship was dead and its spirit lay dormant, waiting for the day it'd be seized by a new generation.

---

He's probably still alive," my wife said one sunny day in the fall of 2019. She had grown tired of me talking about Doug Woolley and his endless contradictions. He was a man savvy but unsuccessful, widely known but with no credentials, full of promise but with little to show for it. He was also an anomaly to track down: everywhere in old clippings and yet nowhere to be found. If he had disappeared, even evidence of his disappearance had vanished, leaving me lying awake at night wondering if Woolley ever existed at all.

I knew he had. I assumed that Woolley had met the fate of almost every other wreck hunter from the early era of wreck hunting, the mid-twentieth century, when bursts in technology and the end of World War II opened the oceans to those hungry to explore. Many didn't even need to leave home, just to revel in the rush of headlines announcing that someone on a boat somewhere had found something. A new species of whale, a new undersea volcano, or, best of all, an old ship from the Romans, the Vikings, or a Spanish galleon filled with gold. And, of course, any news about the most famous wreck of all. Obsession has been known to eat a man alive, so having seen no trace of Woolley for the past decade, I came to the natural conclusion that if he didn't finish his quest, then his quest finished him.

And then I caught a lead. While digging one day through a newspaper archive, I found an old article from 1998 that announced Woolley had written a short autobiography called *One Man's Dream*. Like everything in his life, Woolley produced, edited, and marketed the book himself with hardly any help and no experience in book publishing. And like everything else I knew about Woolley, he had better luck than one would expect. At the bottom of the article, he asked the newspaper to include his home address and urged readers to send him a check for £10 if they wanted a copy. That was the only way he marketed the book.

I looked up the address and saw there was a store next door that sold auto parts. I called the store and asked a thoroughly confused British auto mechanic if an old man named Doug Woolley still lived nearby.

"You mean the ship guy?" he said in a thick accent.

I was quiet a moment and said I'd been trying to get in touch with him and could he deliver a message?

"Call me back tomorrow," he said.

When I did, I got another guy on the phone, and he offered to go to Woolley's apartment and knock on the door. He had some success locating Woolley because he told me to call another guy, who told me to call another guy. Five people later I was texting with a guy named Gary, who identified himself as Woolley's "associate." Going through so many layers seemed like what you'd expect if you wanted to speak to the pope. It reminded me of an old trick a friend of mine once used to secure a dinner reservation: he called a restaurant pretending to be his own assistant.

"We've looked into you," Gary told me the first time we talked. "Doug wants to meet you. Can you come to London?"

As a reporter, I had interviewed politicians and actors. I had investigated sensitive criminal cases and once even met a source in a garage. But something about this made me feel nervous, even intimidated. Doug Woolley had gatekeepers, many layers of them, and I was invited behind the curtain. Was this the seasoned technique of a man building mystery and intrigue with sleight of hand and perceived exclusivity? Or were all these people the infrastructure of an older person who needed extra help?

I booked a flight to the UK. Weeks later, I would realize that my timing had been extremely lucky, both in the trajectory of Woolley's long and roundabout life and amid the world heading into a public health crisis. Even stranger, I would discover that not only was I looking for Woolley, but in a way, he was looking for me, too.

Chapter 2

# THE DEATH AND BIRTH OF GREAT SHIPS

Even in the world of shipwrecks, where there is no shortage of buffs, obsessives, and wackadoodles, Doug Woolley was in a class of his own.

Few things mattered deeply to Woolley, who, as a boy, received repeated lessons that there was nothing notable about the world or even his place in it. One summer when he was a boy, Woolley spent weeks making a model of a double-decker bus out of ice cream lids to be displayed at a nearby festival. On the day of the festival, Woolley came down with a case of pneumonia and was in the hospital. The bus won several awards and appeared in the newspaper, but Woolley had been erased as the maker of it. His mother, uninterested in consolation prizes or encouragement, told Woolley that his luck was consistently and irreparably bad. "At this rate, you'll never get your name in the newspaper," she told him. She saw her son as a boy of eccentric demeanor and interests, destined for a life in obscurity.

A parent's lack of belief in one's potential would be debilitating for most children. But for Woolley, it pushed him into an elaborate inner world where he pulled the strings and no one could convince him he was wrong. The way young girls played with dolls, Woolley fiddled with figurines of soldiers and ships, staging elaborate battles that progressed with intricate dialogue of long-held relationships and grudges. "The characters I created were my best friends," Woolley later said, content with his central place in his own imagination.

A bleak turn of events was a stroke of luck for Woolley. As Britain fought in World War II, Doug's mother was sent to work in a munitions factory, which left Woolley to live with his grandfather, George Woolley. George, also known as Pops, was magic, Doug told people, part mad scientist and part magician who knew the formula to enchant a child. Stuck inside for much of the war that threatened the buildings around Shropshire, where they lived, he listened as George told him tales of Rumpelstiltskin and Cock of the North. One story especially moved Doug. Most people had heard of the *Titanic*, but George lit Doug's imagination aflame by divulging that two of George's sisters, Woolley's great-aunts Sally and Ellen Woolley, had bought tickets to sail on the ship bound for New York. Before Doug could wonder about the fate of his relatives, his grandfather snapped his fingers as if to say, "But wait!" The night before they were set to leave, both aunts had the same dream about disaster striking the ship, and based on their premonition alone, they decided to skip the sailing.

In a final flourish, George said their last-minute change of mind came too late to retrieve their trunks, which were already loaded in the ship's hold. The women escaped with their lives, but nearly all of their earthly belongings, including their valuable jewelry, now sat on the floor of the Atlantic. How they recovered and where they went next never came up, and so Doug never knew.

By the time Woolley turned twelve, this boyhood fascination with the *Titanic* had extended to all ships, old and new. The stories could wind Woolley into a frenzy so deep he might've drowned in it. By 1955, his passion for sunken ships was so strong that any nautical document, photograph, book, or captain's hat that he could get his hands on, he kept as though they were part of a grand web of conspiracy. He was engrossed not with the routine fact that ships sank but with *how* they sank and what happened to them in the endless abyss of Davy Jones's locker, the mythological spirit of the deep believed to preside over the remains of sunken ships and dead sailors.

Woolley had trouble shaking the details of the event. Here was an irrefutable link that infused his own blood into one of the great seismic events of history. In the autopsy of how a lifelong obsession begins, he was hooked early. He may have been born two decades too late to witness what he considered the defining moment of the twentieth century, but he would witness everything about it still to come, because much of it he would engineer himself.

○ ○ ○

The scramble for lifeboats is one of the main focal points of the *Titanic* story. Those who survived could tell their own tales of desperation, heroics, and searing cold. But those who died gave birth to an early question mark of the incident. There was general mayhem, several survivors reported, and a scramble in the water that grew quieter with time. But beyond that, it was sensitive not to probe any further.

The question of how people behave when faced with existential crisis on a sinking ship was such a mystery that, in 1963, a young British scientist named William Keatinge decided that the best way to find out would simply be to ask. Keatinge was a research fellow at Pembroke College in Oxford and was obsessed with how quickly victims of a shipwreck died when they were stranded in the open ocean. It was a hard thing to study in simulation or to observe in real time, but the perfect opportunity arose two days before Christmas that year when a ship called the TSMS *Lakonia* sank about five hundred miles off the coast of Portugal.

During World War II, the Dutch government used the ship to shuttle Allied soldiers, and afterward sold it to a Greek company that converted the military cabins into luxury staterooms for cruising up and down the eastern Atlantic from England to the Canary Islands near Morocco. Approaching the end of its working life, the *Lakonia* was carrying more than one thousand passengers and crew on its unexpectedly final voyage, a Christmas cruise in 1963. A fire broke out in the hair salon at about ten o'clock in the evening. Most people escaped in lifeboats, but the remaining emergency rafts were damaged by the flames, which left about one hundred fifty people with no choice but to help themselves to a drink at the ship's unattended bar and then jump into the Atlantic.

It was the jumpers who interested Keatinge. Most succumbed to hypothermia, the clinical state when one's body loses heat faster than it can produce it, like a punctured rowboat taking on more water than can be bailed out. Hypothermia happens twenty-five times faster in water than in air, and in cold water it has been known to occur in as little as fifteen minutes.

But miraculously, about two dozen jumpers maintained sufficient body heat for several hours until help arrived. Wasting no time, Keatinge sent questionnaires to the survivors' homes and hospital rooms. He asked them to overlook the insensitive timing of his request for the sake of science and fed their egos by asking for "advice" that they could pass along to future shipwreck victims. The survivors who responded said they swam around in circles to keep warm. Some decided to put on extra clothing before jumping. One man reported he took his pants off but kept his coat on.

Keatinge concluded that those who fared best wore heavy clothes and swam vigorously enough to keep warm but not so much as to exhaust themselves. But beyond these small measures, the ultimate determinant of the fate of the *Lakonia*'s victims (and most shipwreck victims) was not their own ingenuity but the speed of their rescuers. A body will become hypothermic even in warm water given enough time.

Luckily, four ships arrived within five hours to pluck people out of the water. A group of American C-54s flew low over the smoky wreck site and dropped flares, rafts, and food to victims who were spread over several miles. Eventually the lifeboats were picked up and the floating corpses were collected. The last person to be rescued was the *Lakonia*'s captain, who was still clinging to a smoky piece of debris moments before it finally went down.

° ° °

Among all the diversity of shipwrecks on earth, there are several broad categories. The reference point for most wrecks is their sinking date, such as the British merchant ship *British Army* and the Chinese *Tek Sing*, which sank on opposite sides of the planet on the same day: February 6, 1822. Others are separated by purpose. Warships are different from cargo ships, even though, in the case of the Brazilian battleship *Aquidabã* and one of the earliest refrigerated ocean liners, the RMS *Magdalena*, they sit on the same stretch of seabed off the coast of Brazil.

People who study shipwrecks tend to sort them most often by geography. The lion's share of World War I wrecks are in the Atlantic, while World War II wrecks are split between the Atlantic and the Pacific. But you don't have to go far or deep to find one. Most people on earth are at any moment less than one hundred miles from at least one wreck. There are wrecks in every ocean and every lake. There are abandoned wrecks lying in the desert of Namibia and under cornfields in Kansas. The streets of San Francisco are built upon hundreds of wrecks that carried gold prospectors to California and were promptly abandoned. There's even a wreck in the middle of New York City. After the Twin Towers fell on September 11, investigators found under the rubble a shipwreck of mysterious origin. They dated the ship to 1773 by studying the rings in its wood planks and surmised it was probably an old merchant vessel carried on land and turned into fill.

Not all wrecks are worth such forensics. Most aren't. For every mysterious Spanish empire wreck carrying valuables like gold or teakwood, there are thousands of humdrum cargo vessels and old schooners from unsung explorers now appreciated only by fish. Not far from where I live in California, the Santa Barbara Channel has claimed hundreds of exploration ships, cargo vessels, and fishing boats that got stuck on rocks or were jostled by waves or caught fire and sank. Virtually all of them are more interesting as wrecks than they ever were afloat.

Wrecks, however, can be valuable depending on what they carried. Before armored cars or private jets, transferring valuable material between countries or faraway colonies was safest when that material traveled by sea, which explains the allure and risks of piracy. Shipwreck hunters like to throw around the number $60 billion as the value of cargo waiting to be discovered at sea, but any estimate is just a guess, many overblown in a field of one-upmanship and hyperbole.

To date, the most lucrative shipwreck ever found is the Spanish ship *San José*, a galleon sunk in battle by the British in 1708 during the War of Spanish Succession. At the time, the ship was traveling back to Spain from Colombia carrying gold, silver, and other valuable cargo like indigo, leather, and precious woods, which all sat undiscovered until 2015. Upon hearing the news of the ship's unearthing, the Colombian government claimed the wreck and its cargo as its heritage. This was accurate, considering the items had come from Colombia. It was also lucrative, since the high price of silver valued the goods as high as $17 billion. Giddy to be presiding over such incredible good fortune, the president of Colombia called it "the most valuable treasure that has been found in the history of humanity."

Chasing gold or silver is timeless. But ships themselves also have value. Depending on the era when it sailed, a vessel's hull is made of materials that may be worth the immense work required to pull it up. Even in bad condition, gleaned steel is worth a few hundred dollars per ton. A steel-hulled tugboat wouldn't be worth the effort, but a large destroyer dismantled by divers could fetch a few hundred thousand dollars. The older the steel the higher the price, especially if it was produced before the nuclear age, when modest levels of atmospheric radiation began to be melted into new steel, making it slightly weaker.

Professional salvage companies don't get out of bed for steel scrap worth two or three hundred thousand dollars. But pirates do, along with poor countries that need the money. In 2016, divers working for the Dutch government descended more than two hundred feet in the Java Sea to survey the wreckage of two vessels, the HNLMS *Java* and the HNLMS *De Ruyter*, both sunk during World War II in the Battle of the Java Sea. When the divers got to the seabed, in place of one of the ships that had been the length of two football fields, they found a crater. It was last seen in 2001, but in the years since, scavengers had illegally removed every ounce of steel, down to the screws. Whoever took it worked quickly. Not far away, they found similar impressions where a three-hundred-foot submarine, the USS *Perch*, had been. Two British ships were also missing, the HMS *Encounter* and the 574-foot *Exeter*. Their survey of eight ships came up nearly empty-handed. All except for one: the USS *Houston*, a cruiser that was subjected to four torpedoes in 1942 and took seven hundred lives when it sank, was still there. But dozens of rivets were already missing.

The *Titanic* will never suffer this indignity. Its depth alone is a shield from poachers. To even get two miles deep requires technology available only to militaries and the richest of the rich. Its forty-six thousand tons of steel—a gargantuan amount in its day but barely half the quantity of modern cruise ships—would hardly be worth the cost of going down and getting it. But that doesn't make it worthless. The *Titanic* carried an abundance of riches, including four cases of opium, five Steinway pianos, the first-class passengers' jewels, and the celebrated 1814 oil painting *La Circassienne au Bain* (valued at more than $2 million in today's dollars). But by far the most valuable component of the *Titanic* shipwreck was the wreck itself, measured in the priceless currency of cultural nostalgia.

° ° °

When I was a kid, I had a hard time understanding how something as big as a stadium could float when something as puny as me required swimming lessons not to sink.

When I was in first grade, my dad thought the basic principles of floating and sinking would make a good science project, so we went to work testing whether a grape, a paper clip, and a bunch of odds and ends around the house would float or sink in water, corn oil, or honey (key finding: almost everything floats in honey). The other kids did simple projects with modeling clay and electric lights, but my modest experiment about liquid displacement somehow struck a nerve. I won first place, thanks entirely to my dad, as science projects tend to go. But honestly, decades later, I still struggle to wrap my head around it.

The beginning and end of shipbuilding comes down to a single number: its displacement. Displacing a quantity of water that weighs more than the ship itself is crucial to floating. If you flipped a cruise ship upside down, water would permeate into the staterooms, libraries, and dining salons, and suddenly, the ship wouldn't be displacing much water. That, more or less, is what happened to the *Titanic.* As water rushed in, had every man, woman, and child bailed out water with their top hats, it might've stayed afloat a tad longer. This was the principle that saved the USS *Cole*, a navy destroyer bombed by al Qaeda terrorists in 2000 while the ship was refueling in Yemen. The attack killed seventeen sailors, but a few fast-thinking survivors began to quickly release fuel to reduce the ship's displacement weight relative to the water. Despite a puncture in its hull more than twice the size of the *Titanic*'s, the *Cole* didn't sink or fall into enemy hands. Medals were awarded, sailors were thanked, and after $250 million in repairs, including five hundred fifty tons of new steel, the ship was recommissioned and put back into active service.

Yet for all its formulas, balance points, and scientific measurements, ship making is more art than science. Unlike cars, which are built in factories in large batches, ships are not suited for construction via cookie cutter, at least not big ones. It's been tried and failed. In June 1917, Woodrow Wilson summoned Henry Ford to the White House and asked him to make ships on the same sort of assembly line he had perfected for automobiles to allow the U.S. to dramatically expand its navy for the ongoing war. "What we want is one type of ship in large numbers," Wilson said. Ford agreed and gave it an earnest effort, producing a batch of destroyers known as Eagle Boats at a factory in Dearborn, Michigan. After producing about a hundred, the process ran into snags. Measurements didn't line up, parts went missing, the perimeter alone exhausted fast-walking workers. A car had hundreds of parts and points of connection, but a ship had thousands.

Whether a ship is ten feet long or a thousand, the process of building is generally the same. Build a frame, fortify it with crossbeams and supports, and then wrap it tightly in wood, steel, or fiberglass to make it watertight. With a good oar and a decent sail, that much would get you across an ocean. But if you had money and time, you could go to work making the topside comfortable with decks, staircases, and indoor cabins with toilets. Throw in carpeting, light fixtures, filet mignon, and a four-piece band, and eventually rich people will line up to ride on your boat instead of someone else's.

In 1909, the year construction began on the *Titanic*, the shipbuilding process went like this. First the ship's backbone, the keel, was laid down. Large ribs were fused to it and then covered with plates of steel. These plates were drilled with holes for rivets, which connected the plates together to make a smooth layer of skin. When the plates were lined up, a white-hot rivet was plucked out of a furnace and placed in the tongs of a catch man, who shuttled that red-hot rivet to another man called the holder-on, whose job it was to place the rivet in the hole while two men, one right-handed and one left-handed, hammered swiftly in alternating whacks. It took three million rivets to build the *Titanic*.

A year later, shipbuilders began switching from iron to steel rivets, which were stronger and could be applied by machines. This had the double benefit of increasing consistency and quality of work, but steel came at a high cost, almost double that of iron, and so the board of Harland & Wolff Shipbuilders, the company contracted to build the *Titanic*, found a compromise. They would use steel rivets for the central hull, which was subjected to the most stress from rolling waves, and wrought-iron rivets for the bow and stern. This was a corner cut, but a defensible one. Cutting another corner, however, the company purchased No. 3 bar iron instead of the top-of-the-line No. 4 bar. In 1999, scientists conducted a metallurgical analysis on rivets found at the *Titanic*'s wreck site and determined that the original rivets were brittle and prone to break under stress.

The debate is rarely whether better rivets could have prevented the sinking after the iceberg strike, but whether stronger joints would have allowed in less water and kept the ship afloat long enough for help to arrive.

These sorts of hypotheticals are the manna of shipwreck obsessives. Time can't be reversed, but the endless life of a ship exists in debates about the what-ifs. What if the rivets were higher quality? What if the iceberg had melted a week prior? What if the captain had heeded warnings to slow down? What if it had been daytime? What if there were enough lifeboats? What-ifs will never bring a ship back. But they will extend the life of a wreck as long as there's one person still to wonder what might've happened if one small thing had been different.

° ° °

The notion of oceangoing for pleasure was a distinct novelty in the years before the *Titanic*. The sea was too rocky, too uncertain. Depending on weather, currents, and the shortcomings of celestial navigation, crossing the Atlantic could take two weeks or two months. It was this nightmare of endless bobbing that led the British writer Samuel Johnson to remark, "Being in a ship is like being in jail, with the chance of being drowned."

The triumphant ocean crossings of long-ago explorers like the Vikings, Amerigo Vespucci, and Vasco da Gama had given ocean travel a facade of simplicity, as though all that was required to cross an ocean was an ounce of bravery and the desire to explore. After all, the great explorers went far and wide and most returned to tell about it. But for the average person who traveled, tempting fate on months-long voyages almost always ended in death.

The most tragic of all ocean crossings were undoubtedly those that carried the enslaved. Kidnapped, sold, and often shackled, the overland journeys to ports in West Africa starting in the sixteenth century were brutal enough, and that was before the punishing sea. The earliest slave ships were outdated merchant vessels, many designed to carry ten enslaved people but routinely loaded with forty. As boats grew bigger over time, head counts followed. In 1797, the *Parr*, a 550-ton ship, set off from Bonny Island, Nigeria, with one hundred crew members and nearly seven hundred enslaved people, many laid side to side and some "spoonways," giving each as little as three square feet for the months-long voyage.

Once on the open ocean, conditions on ships were a battle against squalor, disease, and poor ventilation. The dank lower decks of some ships were barely four feet tall, preventing men from standing, and on ships without portholes, the lingering smell of feces and vomit drove men mad to the point of plotting insurrections or jumping overboard in suicide. Those with the mental stamina to go on fought the daily risk of catching malaria and yellow fever, which wafted through many ships, threatening slaves and crew alike. An anonymously published account by a former slave trader in 1884 reported, "So close and foul was the stench . . . they have been known to be put down in the hold strong and healthy at night and have been dead in the morning."

The capitalist enterprise of chattel slavery prioritized profit and, by extension, speed. The ocean was unkind to ships that had been built on bare-bones budgets, overloaded with passengers, and driven long past their normal operational lives. The stress of managing a boat and its cargo led white crewmen—often conscripted into ocean service themselves for unpaid debts—to intensify their already extreme racial cruelty, inflicting whippings, beatings, and the denial of food and water, sometimes for weeks. So common was the practice of racial inhumanity on the high seas that it barely registered with British elites in 1781, when the white crew of the slave ship *Zong*, upon realizing the boat was running low on drinking water, shockingly threw overboard as many as 130 Africans.

There are few records of this era, particularly from the perspective of slaves, most of whom could neither write nor had the means to do so. But based on archives from business transactions and insurance claims, historians generally believe that over four centuries, from 1500 to 1866, thousands of ships brought more than twelve million Africans to ports in the western hemisphere. As many as one in six of them died at sea. Only recently have countries that participated in such atrocities confronted their role in the slave trade and sought to investigate its human toll. In the absence of wealthy funders or treasure hunters to investigate such wrecks, the task has fallen to altruists and historians, a group of whom formed in 2008 a research initiative under the Smithsonian Institution called the Slave Wrecks Project. Every so often, researchers dive to shallow-water wrecks in hopes of illuminating a tragically forgotten chapter in time.

While the sea was colorblind, governments were far kinder to the struggles of white travelers. In the years after the Revolutionary War, ships stationed offshore to collect import taxes as part of a branch of government known as the Revenue Cutter Service often came to the rescue of ships in distress. Their rescues were harrowing and often reflected the horrors of the sea, such as in November 1812, when a cutter crew found a brig that had capsized in a storm and could hear voices trapped in the upturned hull. The crew saved eighteen men and one boy. All were nearly dead from hypothermia.

Years later, as immigration picked up from Europe, Americans grew horrified at the death counts printed in the newspapers along with ghastly accounts of the violent ocean. In January 1837, a three-masted sailing barque near New York was sprayed by icy water and wind in a violent storm. People on land could hear shrieks, but no one came to help until the next morning, when agents boarded the lifeless ship and found all 104 passengers embedded in ice and frozen to death. This episode built public support for the idea that ocean security was a form of national security, and the Revenue Cutter Service was enlarged in stature and budget and given a new name as the United States Coast Guard.

Conditions were different for those who had money. But luxury as we would define it today was far off. A stomach filled with courage was required for the earliest steam-assisted crossing of the Atlantic, the *Savannah*, in 1819. Such a novelty was alluring enough to interest President James Monroe in riding the *Savannah* up and down the coast of the Carolinas. But once it came time to cross the Atlantic, not a single passenger signed up. Much of the voyage wasn't steam-powered at all: once the ship was out of the harbor, the crew turned off the fuel-intensive and loudly humming steam system and hoisted sails instead. Under wind power, the journey from Savannah, Georgia, to Liverpool, England, took four weeks. By 1836, engineers had cut the mid-Atlantic journey to two weeks. Two years later, when the paddle steamer *Great Western* advertised its maiden voyage from Bristol to New York, enough time had passed without a serious accident that the *Great Western*'s owners built cabins for eighty passengers. All the cabins might have been occupied if a fire on the eve of departure didn't spook everyone. Ultimately, the *Great Western* embarked with seven people.

Nearly all advances in shipping in the nineteenth century—propellers, iron hulls—came from England. Motivated by the fact that most of its worldwide colonies were accessible only by boat, the Crown built a dominant navy that spilled over into private shipbuilding. This spawned the Cunard line, which by 1849 had become the top brand name for passenger travel. A decade later, it was Cunard that had the idea to turn ships into luxury hotels with spacious cabins, steam heat, and ornate designs, which sparked an arms race of elegance with the American-based Collins line, each company working to maximize splendor and speed, along with the most valuable factor of all, passenger confidence.

A steady run of accident-free ocean crossings invited larger and larger ships, culminating in 1858 with the *Great Eastern*, a nineteen-thousand-ton leviathan with space for four thousand passengers. The *Great Eastern* was the *Titanic* before the *Titanic*. There was plush upholstery, teak furniture, and walnut decor. Of particular note was that it was designed by the civil engineer Isambard Kingdom Brunel, the era's Frank Lloyd Wright or Norman Foster, who had designed almost all of England's bridges, tunnels, and railways at the height of its hegemonic dominance. Even though the design principles of building on land are considerably different from those for structures at sea, Brunel's reputation lent the *Great Eastern* so much credibility that when it launched in 1858, critics considered it "practically unsinkable."

And yet, also like the *Titanic*, the *Great Eastern* was beset with problems from the start. It guzzled too much fuel. Laborers died building it. It lost money unless it was full of passengers even though it was exceedingly difficult to convince four thousand Brits to take an expensive one-way luxury vacation all at the same time. It was too bulky to navigate narrow harbors. It turned too slowly. Storms didn't just wick water on the deck; they flooded cabins, soaked carpets, and made the journey wearisome, which didn't help with word-of-mouth marketing. The *Great Eastern*'s biggest draw turned out to be not as a moving ship but as a dockside tourist attraction, where it formally retired in Liverpool in 1886. Ten months later, the ship was sold for scrap and broken apart.

In another time, the deficiencies of the *Great Eastern* might have convinced the next generation of shipbuilders and steamship operators that it was indeed possible to build a ship that was *too* big for optimal safety and operation. But the demise of the world's biggest vessel had the opposite effect. A German investor who had grown his nest egg into a small shipping company called White Star Line thought that the problem with the *Great Eastern* was that it wasn't big enough. Beginning in 1870, the modest White Star Line boomed with more than seventy ships, each more impressive than the one before. White Star ships were shorthand for extravagance, a cultural weight similar to Ritz-Carlton or Louis Vuitton. But history remembers the White Star Line for its failures. The RMS *Atlantic* exploded dockside in Halifax in 1873. The RMS *Republic* collided with another ship in 1909 and sank near Massachusetts. And the HMHS *Britannic*, the largest ship White Star had built at that time, was felled by a German mine in 1916.

None of these disasters had the cultural impact of the *Titanic*, which became the White Star Line's scarlet letter. A million successes in shipping can be eclipsed by a single failure, and after the Great Depression hobbled the White Star Line, it never recovered. By the 1950s, what had once been an industry-leading brand name of opulence had been absorbed by a rival and sunk to the basin of history. Painted as the embodiment of corporate folly, it collapsed, broke down, and washed away.

ooo

When the *Titanic* hit the seafloor, those lucky enough to still be floating in lifeboats on the surface recalled horrifying screams. The sensual trauma of witnessing the breaking of steel, the death of one's family, and the extreme discomfort of freezing air would overwhelm even the most resilient amygdala. Strangely, in the years that would follow the disaster, survivors who had been picked up by the *Carpathia* began to say that while they bobbed waiting for rescue, they heard the crash as the sunken ship hit bottom.

Perhaps they did. But in all likelihood, this was a memory created in response to reliving immense emotional trauma again and again. Sound travels in water, but not well, and certainly not upward over more than two miles. Water is denser than air, which has a muting effect on any sound. Anyone who has ever scuba dived knows how difficult it is to convey a noise of any kind to one's dive buddy several feet away, let alone up through millions of gallons of fast-moving current.

But it is possible to hear in the deep sea, and sometimes at astonishingly long distances. Lower-pitched sounds travel with longer wavelengths, which explains why whales are particularly good communicators underwater. This trait evolved over millions of years as whales became social creatures, but even today, their sounds have to be extremely loud to be heard by other whales. The speaking voice of a normal person is about 60 decibels. A passing siren is 120, and a set of firecrackers nearby can reach up to 150. Blue whale calls start at 180 decibels, roughly the volume of a jet engine during take-off. These sounds can last twenty seconds or more, which requires enormous energy. The only louder known sound in the ocean is that made by the sperm whale, whose short click can top 200 decibels. In the unlikely event a sperm whale walked into a bar, one click would temporarily deafen everyone inside.

The *Titanic* didn't make that much noise on impact, and if the survivors on the surface could not hear the elongated calls of whales swimming in the waters below, then there's little chance they heard the crash of the *Titanic*'s wreck as it made contact with the muddy seafloor. Because in addition to the muffling effect of water, sound from the deep sea doesn't travel upward. In the 1940s, marine scientists noticed this phenomenon by testing small explosives that could signal distress for ships hundreds of miles away. The explosives themselves weren't loud enough to be picked up by neighboring ships, but the sound of their explosion seemed to travel farther the deeper they exploded. This was a mystery: Water is water, so how can sound discriminate based on depth?

This would yield another advance in marine research thanks to the needs of ships, and particularly ships rescuing downed pilots in distress. The top one-thousand-foot layer of the ocean doesn't carry sound well because its temperature gradient is too sudden to carry sound waves. The bottom few thousand feet are too compressed for much sound to escape. But the middle layer of ocean is both cold and relatively unsalty, which makes it more hospitable to traveling sound waves. During World War II, a pioneering geophysicist named Maurice Ewing suggested that the U.S. military could take advantage of this strange middle layer of the ocean by placing small metal spheres loaded with TNT in pilots' emergency kits. If a pilot was in distress, he could release the sphere, which would sink to the middle layer of the ocean and explode when it reached pressures at seven hundred fifty feet. The boom would be heard by microphones on coastlines hundreds of miles away that could triangulate the pilot's position. In an era when radio messages could be easily overheard, exploiting ocean hydrodynamics for secret communication amounted to advanced spycraft.

After the war, this middle layer of ocean water became known as the deep-sea channel, or in scientific circles, the sound fixing and ranging (or SOFAR) channel. As time went on, researchers wondered about other ways to use this odd hydro-phenomenon. They placed sound receivers in the deep-ocean channel, often suspended by floating buoys to prevent them from sinking to the bottom, and listened. They could hear whale calls from more than a thousand miles away, suggesting that whales were well aware of the deep-sea channel long ago and likely evolved into deep divers specifically to use it. During the Cold War, militaries listened to conspicuous receivers in the channel for secret submarines that might not otherwise be detected. Eventually researchers built a vast network of sound responders throughout the Pacific Ocean to detect undersea earthquakes and forecast tsunamis.

And yet, despite the cacophonous sounds of the ocean, the *Titanic* in its first moments as a shipwreck sat shrouded in silence. No one would scream from the bottom of the ocean. No ambulances or fire trucks would ever respond to it. The bodies had already cooled and the hull had already broken. It might have been an hour after it touched down or a year, but it didn't matter in these early moments of the planet's newest shipwreck. The damage was swift and could not be undone.

Chapter 3

# THE MOVEMENT FROM ORDER TO CHAOS

In all of human history, through the Renaissance, the industrial revolution, and the space age, we've never settled on a way to explain one of the biggest geographic mysteries of all time: the origin of the ocean. How did a ball of rock formed billions of years ago give birth to a fertile wet planet capable of sustaining life?

The best we've done is to narrow it down to two theories. The first was born from a lack of imagination. Four and a half million years ago, earth was covered in hot magma that would have instantly boiled off any water. Therefore, if water was absent when earth formed, the only plausible explanation is that the planet's oceans had to come *to* earth via icy comets, asteroids, and other chunks of rocks during a period astronomers call the Late Heavy Bombardment. For three hundred million years, icy asteroids collectively holding several million gallons of water slammed into earth. Their rocks broke apart, and their icy deposits melted and filled in the lowest elevations on the planet. Eventually, after the solar system cleared most of this icy debris, the comets stopped, and all the water that had come to earth would be all the water that would ever be on earth.

But there are big holes in this theory, namely, that, contrary to everything taught in high school chemistry, not all $H_2O$ is the same. Some water contains hydrogen atoms with one proton, and hydrogen in other water molecules contains one proton *and* one neutron, a minuscule discrepancy that adds up to what scientists call a different "isotopic signature." Comet water is different from most ocean water—or at least enough of it to rule out rocky collisions as the sole water benefactor.

The second theory, however, fills in these gaps and crevices. If water that flooded the oceans didn't get delivered to earth, then it must have existed inside earth all along. I dissected this theory with Linda Elkins-Tanton, a planetary scientist at Arizona State University who has spent much of her career investigating the birth of the oceans.

"I *love* talking about this," she burst out. "No one ever thinks it's important, but it's *super* important."

There's an astounding amount of water in ordinary rock, she explained, more than you'd ever think. Inside every rock on earth, even the driest grains of sand in the Sahara Desert, are tiny cells of trapped moisture that add up to a lot. A dry slab of solid granite is 2 percent water, which means a cubic yard of granite, a block about the size of a refrigerator, holds *fourteen gallons* of water. Other rocks like clay or mica have even more water, as much as 13 percent of their mass, which means the same refrigerator-size block holds hundreds of gallons. Extracting all that moisture takes enormous energy, but it's in there, just as it was inside the earth when the planet formed. Left alone for millions of years, the water in earth's magma did what water does: sought out a lower-pressure environment and seeped slowly toward the earth's surface, where it turned into steam and formed into clouds.

Then, as the Bible says, came the rains. It rained for somewhere between thousands and millions of years, off and on, year after year. Eventually it stopped, but it didn't stop because the earth's interior had wrung itself out. It stopped because the planet reached a physical and hydrological equilibrium and started a primitive water cycle in which the water would slowly evaporate toward the sun, get caught in the atmosphere, and rain down again. By conservative estimates, every molecule of water that's ever been drunk, peed out, or swum through has fallen as rain billions of times before.

Based on the isotopic signatures of the oceans, the water-from-rock theory has emerged as the dominant explanation for where water originated. Accepting it also comes with colorful tales of other planets as well. If water came from inside rock, then all rocky planets, including Mercury, Venus, and Mars, once had oceans. Why did earth keep its ocean while the other three didn't? Mercury and Venus likely boiled theirs off in the infernal sun. Mars is thought to lack the necessary mass and gravity to hold water. Earth is the miraculous home to the perfect conditions to sustain an ocean and life and, by extension, sail a ship.

Still, all of our planetary water is not as much water as you'd think. If you picture the earth as the size of a basketball, all of the water on earth—in oceans, rivers, lakes, and ice caps—would barely fill up a marble. Freshwater alone would be even smaller, not much more than a grain of rice. The reason the oceans seem so vast is that they're shallow. The majority of the ocean is less than two miles deep, which sounds deep when you consider doing a handstand, but it's less deep when you consider that the only other ocean in our solar system, the mostly frozen sea on Jupiter's moon Europa, is thought to be as much as one hundred miles deep. And even that's nothing compared to earth's diameter, from one side straight through the core to the other, which is nearly eight thousand miles. Two miles is barely the dust that collects on an unused basketball.

That shallowness, though, is the final stroke of luck for life on earth, because if water really did come from the rock inside the earth's crust and emerge as steam and rain down, then that means the water in today's limitless-seeming oceans is barely a trace of the amount of water still inside the earth's mantle and core. A precise quantity is extraordinarily difficult to measure, but by several guesses, there may be ten more oceans' worth of water still inside the earth in tiny cells of rock. One extra ocean on top of our current one would drown nearly everything and everyone. Ten extra oceans would stretch miles into the sky, sloshing their way around an entirely blue planet. A single ocean turns out to be the very most that earth can handle.

Shipwrecks, however, are not simply ocean crafts that sailed on our planet's wet surface until they had a very bad day. Thinking of water as one lifeless body ignores its evolution as a practical force of its own and our complicated relationship with it. When asked to describe what the ocean is, many of us would answer that it's a large quantity of salty water that moves around the world, occasionally crashes into coastlines, provides wild fish for food, and powers the sport of surfing. That much is true, but it ignores the endless layers of life, evolution, and energy that govern everything from sailing to skydiving.

What does any of this have to do with shipwrecks? Imagine if a single country were as powerful as the earth's oceans. No one could expect such a power to be ho-hum and benevolent—and neither is our planet's water. A shipwreck sits in the same class as a forest fire and a tornado, a flexing of earthly confidence and planetary strength. For every time a ship or its occupants exceed the bounds of arrogance, the ocean has a way of toppling tyrants, vanquishing egos, and resetting the board.

° ° °

Senator William Alden Smith was prepared to act quickly. On account of the conflicting jurisdiction of the *Titanic*—namely, that it was built in Northern Ireland, left from an English port, was headed for America, and carried people of more than two dozen nationalities—there was international confusion over whose job it was to litigate the disaster. Smith, the senior senator from Michigan, knew that the vast American outrage and insatiable hunger for details would require Congress to do something. Blame would have to be assigned, prosecutions might be warranted, and civil liabilities might be due to victims' families.

But first would have to come a common set of facts. The lack of a single confirmable narrative in the two days after the sinking had given way to abundant innuendo and rumors. Some reports had the ship still afloat and being towed to Canada, while others reported the *Titanic* had escaped disaster and that there weren't any victims at all. It hadn't helped that the *Carpathia* had waited nearly a full day after picking up *Titanic* victims to telegraph to American authorities a list of survivors.

Senator Smith was less interested in the gossip than in getting as many witnesses as he could corral into a room as soon as possible to compile an account of what had happened. His effort was complicated by the technicality that non-American crew and passengers had no obligation to speak to an American senator. Making matters worse, on April 18, three days after the sinking, Smith received a message intercepted by the U.S. Navy stating that Joseph Bruce Ismay, the chairman of the White Star Line, who had been aboard the *Titanic* and happened to survive, had told several people that he planned to return immediately to Britain without setting foot on American soil.

Smith saw Ismay as a kingpin, a man intimately familiar with how the *Titanic* was built and how it failed, and who might also be liable for shortcuts in design and operation that resulted in such significant loss of life. Smith could compel Ismay to testify before the Senate only if he was in America, and, sensing the urgency of the situation, Smith assembled several of his fellow senators and booked the next train for New York, which arrived barely an hour before the *Carpathia* pulled into Cunard's Pier 54. Smith stood on the dock flanked by two U.S. marshals. When he spotted Ismay, a man freshly in shock from witnessing mass death, searing cold, and corporate failure, Smith pushed through the crowd and handed him a subpoena.

Ismay agreed to testify. But the awkward timing, coming straight from a maritime disaster with throngs of pressmen lusting for details, ruled out a detour to Washington for a conventional hearing in the Capitol.

Agreeing with Ismay's assessment, Smith booked the East Room at the Waldorf-Astoria at the corner of Fifth Avenue and Thirty-Third Street to open the hearings the next morning. Smith hadn't intended any discomfort in booking the hotel and may not have realized that the Waldorf-Astoria had once been the home of American millionaire William Backhouse Astor, whose grandson John Jacob Astor IV, known widely as the richest man in America, had died three days earlier at sea—aboard the *Titanic.*

Less than one hundred hours after he watched more than one thousand people die, Ismay addressed the room of U.S. senators. "In the first place, I would like to express my sincere grief at this deplorable catastrophe," Ismay said. He dispassionately described the way the ship had been built and its ports of call before steaming for the open Atlantic. His testimony at times took the form of a shipbuilding seminar, instructing politicians accustomed to marble halls and hearing rooms on how a steamer actually functioned. The sundeck was on top, he explained, followed by the A deck and the B deck. He explained the location of the bridge, the so-called navigational center of a ship. He relived the day leading up to the collision and recalled his final conversations with the captain, whom Ismay said was unmoved by warnings of ice ahead. According to Ismay, after the collision the captain had quietly ordered the lifeboats to be filled and for women and children to be given priority.

Senator Smith, however, was far from dispassionate. His questions were filled with outrage and hysterics, his expressive face moving between red-hot anger and cool warmth. Smith was officially a Republican, but not one fitting a partisan script. He was a self-described "maverick" and the leader of his own party, drawn to and driven by causes most saw as hopeless, including racial equality and women's rights. A former lawyer and businessman who had become a dismissible populist, Smith was now the public face of the biggest news story on the planet, and as long as he was, he would wring every drop of public attention in the classic style of the grandstanding politician.

Smith started with the most pointed question of all—how Ismay, a man of immense privilege and power aboard the steamer, had arranged his own survival when almost all other men on board had died.

SENATOR SMITH: *What were the circumstances, Mr. Ismay, of your departure from the ship?*

MR. ISMAY: *In what way?*

SENATOR SMITH: *Did the last boat that you went on leave the ship from some point near where you were?*

MR. ISMAY: *I was immediately opposite the lifeboat when she left.*

SENATOR SMITH: *Immediately opposite?*

MR. ISMAY: *Yes.*

SENATOR SMITH: *What were the circumstances of your departure from the ship? I ask merely that—*

MR. ISMAY: *The boat was there. There was a certain number of men in the boat, and the officer called out asking if there were any more women, and there was no response, and there were no passengers left on the deck.*

SENATOR SMITH: *There were no passengers on the deck?*

MR. ISMAY: *No, sir; and as the boat was in the act of being lowered away, I got into it.*

Smith tried to channel the public bloodlust for an explanation about how so many people could have died so horrifically while the corporate boss was sitting comfortably in a New York hotel drinking tea with members of the U.S. Senate. This outrage was familiar to Smith. For decades in Congress, Smith had sought to hold accountable railroad bosses whose businesses served the public interest but were known to pinch pennies and cut corners that endangered the lives of riders. Smith believed the steamer lines behaved the same way and was prepared to craft new regulations for a sloppy and dangerous industry. But to do so required confirming his presumption of negligence. And ideally, he would need to hear from not just White Star officials like Ismay but passengers too, who had little loyalty to a steamer company whose failure had just killed their loved ones and left them freezing for hours at sea.

Nine witnesses told their stories during the hearing's first two days, not yet a week after the disaster. On the third day, Smith moved the proceedings back to Washington, having tired of the incessant interruptions from eager reporters, desperate family members, and riled-up members of the public. Back in the Capitol, Smith continued the questioning for sixteen more days, calling a total of eighty-six witnesses and building a transcript file nearly twelve hundred pages long that would be the closest thing the *Titanic* would have to a contemporary eyewitness account. The British Board of Trade conducted a separate investigation. Together, the findings from both probes built the foundation of knowledge of the disaster that would inform more than a century's worth of stories, retellings, books, films, musicals, documentaries, poems, songs, video games, and endless interpretations, including one from a British composer in 1969 who created a "soundscape" exhibition that played recordings of survivors' testimony and emergency Morse code messages. The idea, he said, was that the story "never completely dies but merely grows fainter and fainter."

The investigation had a practical effect as well. With his closing speech, Smith submitted a list of proposed regulations for all ships that passed through American waters. They should slow down in areas of ice and create a firm chain of command for captains to be clearly informed of navigational threats. No longer should ship telegraphs be used for frivolous messages between passengers of different ships, and the use of rockets must never be celebratory, only to signal emergency. And most of all, all future ships should be equipped with enough lifeboats for every single passenger.

"This should be the occasion for a new birth of vigilance," Smith said while delivering a set of reforms to the shipping industry that would become his single greatest legislative contribution over twenty-four years in public office. "Future generations must accord to this event a crowning motive for better things."

° ° °

Defining a shipwreck is a bit like defining a car crash. Two components are required: a car and a moment of impact. But from there, endless variables obscure simple explanations. You might say a wreck is a watercraft that runs into trouble at sea. But trouble isn't the same as sinking, nor is the ocean the exclusive keeper of wrecks. Is an accident required, such as the *Titanic*'s swipe with an iceberg? Perhaps not when an intentional torpedo can dispatch passenger liners like the RMS *Lusitania* or the SS *Athenia*. Strangely, a shipwreck doesn't even need to be a ship; crashed airliners, flooded trailers, and Jeeps carried out to sea are wrecks all the same. Summing up a wreck may be less a matter of physics than of philosophy. According to the colonial scholar José Rabasa, a shipwreck's central quality is simply that "it marks the movement from order to chaos."

Usually, the deeper the fall, the harder the crash. Any falling object reaches its terminal velocity when the dragging force of resistance equals the downward pull of gravity, which explains why a single glass marble would sink nearly ten times faster than a ship as large as the *Titanic*. Had the engineers who built the *Titanic* wanted the ship to sink as fast as possible, they might have made each surface smoothed and rounded, a bit like a torpedo. But sinking speed would be a poor priority. And anyone who's spent months living on a dank submarine would tell you that such a ship would be dreadful to sail on.

A ship is in danger the moment it starts taking on more water than it can bail out. There are exceptions, of course. Puncture your canoe and you'd have about thirty seconds. Ships as large as the *Titanic* can take a few hours. If sinking speed were a race, in dead last would be the SS *Atlantus*, an old concrete ship that's been sinking since 1926 off the coast of Cape May, New Jersey, slowly receding into shallow water every year. Concrete ships were a wartime experiment, ordered into production by Woodrow Wilson. But their design made little room for cargo, which left them expensive to operate. Since then, the *Atlantus* has had a tough time, exposed to both sun *and* salt water, two of the most corrosive forces on the planet. But being half-sunk has also given it new life. Plants grow on it, birds build nests on it, and at one point someone even put a billboard on it.

There are scientists who study the ways ships sink. The field of maritime forensics inspects how ships are built and the variables in construction, cargo, and navigation that give some vessels a far better chance than others of surviving an accident. Parks Stephenson, a marine forensic analyst, has been studying the phenomenon known as shipfall for decades. Most of all, he studies the precise position of the *Titanic* as it sank.

"I've been trying to get away from it for years, but it keeps pulling me back," he joked, without really joking.

Stephenson has looked at so much data for so many years that he no longer believes that the *Titanic* sideswiped an iceberg, an assertion that's gotten him literally laughed out of rooms. Based on the firsthand accounts and the position of debris on the seafloor, Stephenson believes the ship grazed the iceberg from below, a motion known as grounding that's consistent with the intense vibrations survivors recalled, rather than being jolted across a room, as might be expected from a strike on one side or the other. Nevertheless, the sideswipe explanation remains the consensus of researchers.

Every ship sinks differently. Some bow first, others from the stern, and still others from a capsizing roll. Yet the two measures of physics that will explain how any ship falls are buoyancy and equalization. When a formerly buoyant ship sinks, all remaining pockets of buoyant air will move upward in the ship, sometimes flipping the ship over.

Eventually, when all buoyancy has been released, the vessel is referred to as equalized; internal water pressure is equal to external water pressure. At that point, any ship, no matter how big, small, old, or young, a high-end steamship or a working-class lobster trawler, will succumb to weight distribution and sink with its heaviest side down. Often, the heaviest components, like engines and boilers, are in the hull, which explains why ships tend to settle upright. But there are the occasional anomalies when ships sink topside down, usually owing to heavy cargo carried higher in the hull. Wreck divers flock regularly to the USS *San Diego*, about ten miles from Fire Island, New York, to see a World War I navy cruiser that was dragged down by its top-heavy weapons systems and has spent the last century sitting awkwardly upside down.

In the same way, any water can take wrecks: shallow, deep, fresh, salty, icy, or warm. You could look at the high density of wrecks in the English Channel or the coastal waters of Spain and posit that England and Spain have the world's deadliest seas. But the reason is more likely that the Spanish and Brits sent out more ships between the sixteenth and nineteenth centuries than almost all other countries combined. This also explains the mystery of the Bermuda Triangle, which turns out to not be much of a mystery. Several years ago, an Australian scientist named Karl Kruszelnicki published a study that showed that the high concentration of ships that run into trouble in the triangle between Bermuda, South Florida, and Puerto Rico is not the result of aliens, crystals, or hydromagnetic whirlpools. "It is close to the Equator, near a wealthy part of the world, America, therefore you have a lot of traffic," Kruszelnicki said. Sitting directly in the path of Atlantic hurricanes doesn't help, either. Studying the numbers, he concluded that vessels wreck in the Bermuda Triangle at the same rate as anywhere else. A historian with the Naval Historical Foundation put it more bluntly: "To say quite a few ships and airplanes have gone down there is like saying there are an awful lot of car accidents on the New Jersey Turnpike—surprise, surprise."

(The U.S. Coast Guard has come to the same conclusion about the Triangle, which all but proves a government conspiracy.)

If the waters east of Florida really are a nothing-to-see-here patch of ordinary ocean, how can anyone explain the ships that disappeared for no reason, like the USS *Cyclops* and USS *Proteus*, two navy cargo ships that vanished in 1918 (*Cyclops*) and 1941 (*Proteus*) without a trace of evidence while in the Bermuda waters?

Science has a strange answer for this too, and it has to do with water. The density of a ship is fixed. But if you reduce the density of the seawater, such as with gas bubbles from deep-sea methane vents known to exist near Bermuda, a ship will suddenly become *more* dense by comparison and start to sink. Felling a cargo ship would require relentless and voluminous methane, but at least in a lab, scientists have shown it could work. For its part, the navy says the *Cyclops* and *Proteus* were probably hit by storms or mines and sank, simple as that.

Bermuda or anywhere, not all wrecks are accidents. Sometimes humans actively root for wrecks, up to the point when they are forced to create one urgently, the maneuver known as scuttling. Steel-hulled ships are designed to last as little as fifteen years if they're driven hard or up to forty if they're well maintained. But more than that is highly unusual. Eventually, all ships become less efficient, harder to operate, and more expensive to maintain. When their day comes, all ships go to the great shipyard in the sky, which, as it happens, is most often underwater.

Scuttling was once a method of war strategy. Hernán Cortés famously scuttled his entire fleet of ships in 1519 to prevent his crew from deserting him after they landed in Mexico. Following Germany's surrender in 1919, German naval commanders decided to scuttle their entire fleet of fifty-two vessels to prevent the British navy from seizing the ships and using them to attack Germany. The Germans opened watertight doors, disengaged hatch covers, and cleared all portholes to invite in water. The British didn't notice for hours, until the listing of the largest destroyers had become obvious, but by then it was too late. One can imagine the displeasure of the British, who, in the most British of protests, reluctantly rescued the German soldiers who were waiting patiently in lifeboats.

Modern warfare, however, allows the intentional disposing of multimillion-dollar assets only in the most dire of circumstances. More often, ships are scuttled as a way of disposal at the end of their working lives, either to become artificial reefs or occasionally for navy target practice, as was the case of a decommissioned thirty-six-year-old U.S. frigate called the *Ford*. In 2019, the U.S. and Singapore armies cleaned out the *Ford* of all fuel and electrical equipment and pounded the frigate with dozens of Harpoon and Hellfire missiles and then studied the precise way it broke apart, collapsed, and took its leave to the depths of the Pacific.

Cargo is different. Scuttling unwanted or dangerous cargo often requires a ship to sink it in order to ensure it actually sinks and doesn't wash up on a crowded beach somewhere. In the latter half of the 1960s, the U.S. Department of Defense commissioned a quiet but dramatic scuttling mission called Operation CHASE—short for "Cut Holes and Sink 'Em"—to dispose of World War II–era munitions off the coast of Florida and the Bahamas. The rationale behind the operation was that the U.S. military had made a surplus of chemical weapons in case they were needed to fight the Nazis, and, after nearly two decades storing them in warehouses, they decided that the ocean was the best place to ditch them. No member of the U.S. Congress would want chemical weapons buried underground in their district, plus the ocean was so limitless and complex in chemistry that, it was thought, it would somehow neutralize the weapons without any danger or damage. This was a convenient assumption that facilitated the fast dumping of the weapons-filled canisters, often by ship captains who were so anxious to offload such dangerous cargo that they were known to toss them overboard a few miles offshore instead of the requisite one hundred miles or more, where they would sink in much deeper water.

Such reckless conduct might have been met with legal challenges and public protests if Operation CHASE hadn't been classified. The navy didn't want America's adversaries to know they could effectively dumpster dive in international waters for some of the most dangerous biochemicals on earth. In the era of DDT and Rachel Carson's *Silent Spring*, officials also wanted to limit public outrage about ocean pollution and dangers to public health from chemical weapons, which are notoriously dangerous to transport, store, and destroy. But this didn't last long, nor was the navy entirely honest about the extent of the program. In the summer of 2004, a ship dredging for clams off the coast of New Jersey accidentally pulled up an old artillery shell that, when opened, appeared to be filled with black tar. Someone had the sense to call a weapons team, and when three technicians arrived, they all incurred skin burns and pus-filled blisters on their arms. It turned out the shell was filled with concentrated mustard gas.

In fact, seawater does break down most human-made substances—even chemical weapons—but not without extreme danger to marine and human health. When exposed to seawater, mustard gas forms a thick gel that rolls around the ocean floor for as long as five years. It's deadly, but not as deadly as nerve gas, which the navy also admitted to tossing overboard. A few drops of nerve agent can kill a blue whale in a minute. It lasts about six weeks before it breaks down into its nonlethal chemical parts, but in that time, it'll kill everything it touches. In 1987, hundreds of dolphins washed up on beaches in Virginia and New Jersey with extreme burns and blisters that resembled a reaction to mustard gas. Biologists confronted navy officials, who admitted the Operation CHASE dumping of chemical weapons was in fact *slightly* more widespread than they previously said. The navy promised an investigation, but it was short-circuited by Pentagon leaders, who concluded that the dolphins died from a bacteria or viral infection and left it at that.

As it turned out, the program was far more extensive than the American military ever let on. The Pentagon now admits that the dumping program began in 1944 and proceeded for more than twenty-five years. Contrary to claims that the dumping was isolated to international waters off Florida, there were actually at least eleven dump sites, six on the East Coast, two on the Gulf Coast, and a handful in California, Hawaii, and Alaska. The sites aren't all known—records were sparsely kept at the time, perhaps intentionally. But in all, the scuttling operation is believed to have included sixty-four million pounds of nerve and mustard agents; four hundred thousand chemical-filled bombs, land mines, and rockets; and more than five hundred tons of radioactive waste.

All of it was too dangerous to abandon on land, and so it was relegated to the ocean, in equal parts everyone's problem and no one's.

"It's impossible to know the true danger even today," Craig Williams, a chemical weapons destruction expert, told me. "Whether it's gotten more dangerous from corrosion of the canisters or less dangerous from dispersion in the water is anyone's guess. But this stuff is unequivocally dangerous, and it's still out there."

o o o

In the weeks after the *Titanic* disappeared, engineers and laypeople alike debated the condition of the ship as a newborn wreck. Where *was* it? they wondered. And did it sink to the seabed, or was it somewhere else?

On May 16, 1912, a month after the ship disappeared, a writer for the *Medford Patriot* in Medford, Oklahoma, explored the idea long debated by sailors that ships don't sink to the bottom but instead stop halfway down and become suspended in the water column for all time. This was based on the erroneous calculation that with depth, extra pressure resulted in water becoming more dense. Therefore, once the ship crossed the threshold from water less dense than the ship itself to water more dense, it would stop sinking and simply hover, no longer a ship and not yet a wreck. “On account of the weight of water at great depth the density [is] so increased that dead bodies or sunken ships would not go down beyond a certain depth but remain eternally suspended in the sea drifting about at the same depth,” the *Patriot* reported.

In the aftermath of the *Titanic*, this kind of scientific lore was soundly rejected by both the U.S. Hydrographic Office and the editors of *Scientific American*, who wrote in an editorial that, no, water does not become denser with depth. In fact, laboratory experiments at the time had shown it was easier to increase the density of steel than it was to compress the molecules of water, which made the idea of a significantly denser form of liquid water inconceivable. Moreover, government engineers added, the increasing water pressure would have compressed the *Titanic* on its way down, in effect increasing *its* density relative to water and accelerating rather than slowing the velocity of its fall.

That debate was settled. But much less clear in both the scientific and public discourses was exactly what sort of conditions surrounded the *Titanic* in its new deep-sea grave. Older ships like the British steamer SS *Copenhagen*, which sank off the coast of Florida in 1900, were well understood on account of resting in shallow water. A scientist sitting above the wreck could take a fairly accurate survey of the fish and kelp and measure the microbial activity in the water to hypothesize how the ships were changing in their marine environment. But the *Titanic* had gone more than a mile deeper than anyone had ever traveled, and the creatures, conditions, and currents of the deep sea were as unknown as the surface of the moon.

One could expect that the deep sea was cold. In 1847, the American astronomer Matthew Fontaine Maury was one of the first to demystify the movements of currents that carried warm and cool water in a constantly moving conveyor belt of biological circulation. Maury become an ocean loyalist after a stagecoach accident in his youth soured his attitude toward anything that occurred on land, and he learned about the ocean currents by studying the logbooks of old ships, some dating back to the American Revolution. Reading the anecdotal reports from long-dead captains about wind patterns, calm spots, whale migration routes, and dramatic changes in ship speed in certain stretches of ocean, Maury pieced together the first maps that showed a current of cold water moving along the ocean floor from the poles to the equator, where it would then be heated, rise to the surface, and be carried back to the poles. This aligned with the stories of mariners who had long pulled up buckets of ocean mud to cool their jugs of drinking water. The seafloor's uninhabitable frigidity was known, but if deeper meant colder, then it stood to reason that, at a certain point, water would be colder than ice.

The public was enchanted by what this meant for the *Titanic.* Was the seafloor made of fine sand particles that were compressed into solid rock? Or were there giant boulders that washed out from rivers and kept rolling deeper and deeper down the ocean basin until there was nowhere lower to roll? A contemporary of Maury's, the British oceanographer Sir John Murray, took on these questions and conducted laborious samplings of seabeds in the late nineteenth century. He concluded that at the *Titanic*'s depth, the ship must lie in a bed of pelagic fine-grained ooze, oily to the touch and so finely divided that "it would take many hours to settle in a glass of water." Murray believed this mud to be the remnants of millions of years of skeletons, cells, and biological sludge and that more than three-quarters of deposits covering the ocean floor had passed through the alimentary canals of dozens of sea animals. No molecule was wasted in the ocean, down to the finest grain of sand.

In a time before humans could collect biological samples from such depths, the most colorful theories about what lurked below centered around deep-sea fish that had to constantly fight the pull of rising to the surface. If they rose too high, they'd die. "It sometimes happens that a deep sea fish, chasing its prey, gets out of its depth and goes tumbling upward," *The Kansas City Star* told readers desperate for any shred of detail from the ocean abyss after the *Titanic*. "As the pressure is relieved [a fish] swells and bursts and their bodies are often found, torn and mutilated, floating upon the surface. When brought to the surface in the deep sea dredges these fish are always dead: they fall to pieces."

This was mostly fiction. Deep-sea fish have swim bladders the same as shallow-water fish, except instead of filling them with buoyant air that would yank them to the surface, they're full of fats and lipids that are only slightly lighter than water. Research later in the twentieth century showed that instead of spending their life in a feverish struggle against rising to the surface, virtually all deep-sea creatures evolved to constantly expend as little effort as possible.

Small life, however, wasn't as compelling as monstrous life. Tales of eel-like sharks circulated in popular newspapers along with theories of ribbon fish thirty feet long that weighed eight hundred pounds. The most mysterious neighbor of the *Titanic* would be the largest invertebrate on earth, the giant squid. Believed to be a hundred feet long with a sharp head like an arrow, the giant squid had only been observed in death, washing up on beaches to horrified crowds. One that washed up in Newfoundland's Trinity Bay in 1877—a "baby" at merely sixty feet long—led scientists to think adults still alive could be double, or even triple, in length, as long as two hundred feet. The giant squid's deep and dark habitat shrouded the species in such extreme mystery that it wasn't until 2012 that scientists filmed a

giant squid alive for the first time. The trick, they
a submersible remote-operated vehicle with
squid can't see, instead of conventional floodligh
frighten anything accustomed to pitch black.

For those grieving lost family members, it wasn't too big a
to wonder if the giant squid and other oversize bottom-feeders might attack the *Titanic* and rip apart its crinkled hull with bloodlust for the bodies inside. Decades before the disaster, the adventure novelists Jules Verne and William H. G. Kingston filled the public imagination with tales of deep-sea assailants like sharks and whales that hungered for human flesh. A lifeless human wouldn't stand a chance against ferocious ocean predators, even though, when it comes to being disembodied, disemboweled, and decomposing from the inside out, it's the tiny creatures that do the most damage.

None of this turned out to matter. The gravest damage to the *Titanic* had already been done in the first ten minutes after its stern disappeared below the waterline. For it was in those few minutes that the ship transitioned from a sleek facade of steel to a chunk of mangled metal. Despite this occurring in the golden age of travel, and despite this ship as a symbol of technology and luxury, the ocean did what the ocean has always done and began to rip the ship apart.

---
o o o

The ancient Egyptians called her Nu, the water responsible for all of life. The Aztecs called her Huixtocihuatl, a goddess who controlled all salt water. In virtually all ancient traditions and fables, the sea is a feminine force of rebirth and generosity. But every culture that has ever lived off the earth's natural resources knows that the ocean is more than just a nurturing spirit of delicate balance. The uit, famous for their plentiful descriptions of water and ice, had a name for the life-giving force of the ocean that could quickly turn dark. They called her Arnapkapfaaluk, or big bad woman.

As bad as things can get for a boat on the surface, things get worse underwater. Once a ship disappears below the waterline, the sea begins its work like a white blood cell, attacking and dismantling the invader not with huge blows but with millions of tiny paper cuts from microbes, sun damage, and rust. Rust is the worst. The simple chemical collision of iron and oxygen behaves like a living organism, and magnified in its limitless impact, rust is the most ruthless predator on the planet. It's especially unkind to boats. The U.S. Navy spends $3 billion a year battling rust, a price tag as high as $10 million *per ship*.

Entire teams of engineers spend their careers worrying about this sort of corrosion. The American Galvanizers Association runs an all-hours corrosion hotline to help people who are building things make sure they last as long as possible. One day I called it and talked to Alana Fossa, a corrosion engineer who was a little more excited than I expected to talk about how materials break down and the ways humans have devised to fight back.

"Most calls I get are pretty off the wall," Fossa told me. She said she hears regularly from people who make bridges and cars or occasionally the suburban dad who's building a swing set or refinishing his deck with new screws. The worst conundrums of corrosion tend to come from people building boats and piers, who know that in the presence of salt water, they're on borrowed time.

"When you build a ship, if certain measures aren't taken, a structure that's supposed to be around for decades can start to break down in months," Fossa said. She explained that every environment will lead to some breakdown, but the factors affecting boats, like high humidity, gusts of oxygen, and chloride from salt, are all maximized in marine environments. This is why it's important to remove a leisure boat from the water when you're not using it, or for a cruise ship or oil tanker to get its hull repainted often to separate the metal from the water. All of this work forestalling the inevitable goes a long way in explaining the old joke among boat owners about how the best two days of owning a boat are the day you buy it and the day you sell it.

Not all ships break down so quickly. The reason why some shipwrecks disappear in weeks and others last centuries comes down to the basic scientific principle that environment is everything—and given the right conditions, damage can be stalled for thousands of years.

Several years ago, I went to Italy to investigate the life of the famous Italian mummy known as Otzi, who lived five thousand years ago. Otzi had been walking in the Alps one day on the modern-day border of Italy and Switzerland when he was shot by an arrow and killed. His tooth enamel later revealed that he grew up in a nearby village, and the size of his leg bones indicated he was accustomed to long walks in the mountains, perhaps as a shepherd. Five thousand years later, it was a miracle we could know this, let alone that Otzi's body still existed at all. The only reason why is because the place where he was killed stayed cold and dry, the best possible conditions to preserve anything.

There are such conditions in water too, spots of oceans and inland seas that combine the optimal factors where ships break down at a pace so slow it makes glaciers seem hyperactive. Below a certain depth, reduced oxygen slows down the chemical reaction that creates rust. Ships found below ice in the Arctic are relatively intact, as are vessels in the cool and freshwater Great Lakes of North America. The *L.R. Doty*, a Gilded Age cargo steamer that sank in Lake Michigan in 1898, was in remarkably spry shape in 2010 when divers found her and her cargo of corn under three hundred feet of freshwater. This finding and more like it helped birth a new trend in winemaking in California and France, where vintners age bottles underwater to take advantage of reduced oxygen and cool temperatures, more stable than even the best insulated wine cellar. What started as an industry gimmick has been granted credibility in blind tastings, giving birth to the term *aquaoir* for the underwater conditions that give a wine its flavor.

By the same degree, the best wrecks for scuba diving are never the very old ones. For a wreck to be accessible to divers requires it to be shallow, and shallowness brings its own set of problems, including sun, oxygen, and ocean creatures. One of the most popular diving wrecks in the world, the Italian troopship *Umbria* off the coast of Sudan, is in good condition, at least for its relatively young age of being a century old. Every year thousands of divers swim through its bridge to check out its equipment and take underwater pictures with hundreds of wine bottles and more than 360,000 aircraft bombs (look but don't touch). Rusting slowly, however, is still rusting, and eventually the *Umbria* and its bombs will break down until there's nothing left.

If the goal is to keep a defunct ship around for as long as possible, there's no better place to sink than in the Black Sea. In 2016, a team of scientists with the Centre for Maritime Archaeology at the University of Southampton found the oldest shipwreck on earth off the coast of Bulgaria. And when they looked deeper, as deep as six thousand feet, they found a bounty of other ancient Roman- and Greek-era ships, some dating back three thousand years. Even more striking was that most of them were wood, which in virtually all terrestrial conditions will break down faster than steel, hence why shipbuilders abandoned wood hulls for steel ones in the mid-nineteenth century. But the lack of all oxygen in the Black Sea below five hundred feet has allowed organic materials like wood and rope to sit essentially stagnant, preserved with the same ancient carvings and drawings as the day they sailed thousands of years ago.

By contrast, the *Titanic* is a Goldilocks-style wreck: not in the worst conditions, but still not great. There's very little dissolved oxygen at twelve thousand feet, so the *Titanic* has been spared for more than a century from conventional rusting. But in 1986, researchers, using remote and autonomous vehicles, discovered reddish icicle-looking structures growing on the *Titanic*. Several samples were brought to the surface, and when studied under microscopes, they turned out to be enormous deposits of rust-like growths that were aptly named rusticles.

For all the attention to an old boat that started its life as a beacon of modern invention, it's worth pausing to consider how totally it's been felled. First by the solid version of water and then by the salty kind, and then finally by something so small and strange we've lived almost all of human history not knowing it exists. There's something poetic about how something so mighty could be not only so easily demolished, but slowly erased from the planet by the tiniest earthly species that collectively decides what lives and dies.

Chapter 4

# MERELY A MATTER OF MAGNETS

Crossing the ocean has always taken a measure of audacity. Early cultures often set off by necessity, to find fish or scout for a new land. Preparation and experience made the difference between voyages that ended in triumphant homecoming and others that wound down in silence and mystery. More than any other factor, though, the one that seemed to matter most was a sailor's confidence that he could tame the ocean and that everything else would work itself out.

The earliest sailors, like the Polynesians, the Vikings, and the Arabs, demonstrated that success was possible. European empires sent fleets to conquer the other side of the world, fueled by the new technology of three- and four-masted sailing ships combined with navigation technology like the magnetic compass, the sextant, and the astrolabe, which calculated one's position based on the angles of celestial bodies. This not only led to cultural clashes between conquerors like the Spanish, Italians, and British, and their conquered, the Native Americans, the Aztecs, and the Indians, but also made possible new feats—chiefly, circumnavigating the world—that expanded the bounds of what humanity considered nautically possible.

The age of European empires was further proof that the ocean could be used to overpower faraway civilizations that were dependent on the ocean. The Indonesians and Japanese realized that the same force that sustained their lives and culture could also be turned against them by foes with bigger ships and more violent convictions. The domination of those who possessed ocean experience over those who didn't was absolute, up to the point that the world-leading British navy could not only transport human slaves by sea, but land them on continents once filled with *other* native people and force them to work out their differences.

The one constant of seafaring progress, however, was setbacks. In 1782, an entire fleet of British ships was destroyed by a hurricane in the North Atlantic, barely a month after the even more tragic and embarrassing wreck of the HMS *Royal George*. After the American Revolutionary War, the *Royal George* returned to England and was docked near Portsmouth on a day so docile that more than three hundred fifty women and children were invited on board to tour the ship. As the visitors wandered around, the crew set out to repair a water-intake valve three feet below the waterline on the starboard side. Guns and heavy cargo were moved to tip the ship, but it happened too fast, and the shift in weight caused the ship to rock back and forth so forcefully that the *Royal George* began to take on water and then sank. All but a handful of the visitors died, along with five hundred crewmen. Despite the obvious miscalculation by the crew, an investigation of the accident blamed the ship and the "decay of her timbers."

Tragedy continued to strike. In 1820, the Nantucket whaling ship *Essex* was hunting whales in the South Pacific when an agitated sperm whale rammed the bow and sank the ship. The crew escaped into a smaller boat and rowed for more than a month until they pulled up on an outcrop called Henderson in the Pitcairn Islands. Incredibly, had they washed up on a neighboring island, they would have been greeted by fellow English speakers who were the descendants of mutineers from Captain William Bligh's infamous 1789 mutiny aboard the HMS *Bounty*. But on Henderson Island, they succumbed to cannibalism and, in accordance with the so-called Custom of the Sea, drew lots to determine who would be killed to feed the others. The story of the *Essex* became as famous then as the *Titanic* is today, and three decades later, in 1851, its ubiquitous fame inspired a New York author to write a fictional account of the disaster he titled *Moby-Dick*.

Prior to the *Titanic*, one of the worst passenger disasters had been the SS *Arctic*, which, when launched in 1850, *The New York Evening Post* called "the most stupendous vessel ever constructed in the United States, or the world, since the patriarchal days of Noah." The paper foreshadowed that not "a single accident" would befall the ship, which was regrettable four years later when the *Arctic* collided with a French steamer in foggy waters near Newfoundland. The *Arctic* was yet another defeat for shipbuilding hubris, and it again laid bare the fiction that women and children were given first rescue. Crewmen filled the six lifeboats and left two hundred helpless passengers to die. Not wanting to draw attention to these facts unbecoming a great ship in the new era of ship safety, neither the *Arctic*'s owners nor the U.S. government conducted an inquiry into the accident, and the wreck has never been found.

Still, the arc of nautical progress bent toward triumph. The bigger, the stronger, the warships that toppled empires, and the others scuttled in defense. The steamers that shrank the planet for trade and travel, the dinghies and barges that made ocean existence not only safer and smoother but even pleasurable. All because humans had confronted problems posed by the oceans and, by and large, solved them.

This was the streak of human accomplishment that led a man in 1914 to believe that the *Titanic* wasn't permanently gone. The ship had been underwater for two years, a blink in the life-span of most wrecks. To a man named Charles Smith, this meant the ship was in waterlogged but all-around salvageable condition. And that with the right expertise, it could be pulled to the surface, its bodies recovered, its furniture repaired, even the valuable jewelry rumored to be kept in the onboard safes returned to its owners.

Smith was an engineer. He held no experience with oceans or other bodies of water, but he was one of the top engineering minds in Colorado, where he lived. Smith had been born in 1861 at a pivotal time for his young country, and he grew up in a family that held a genetic penchant for oversize earthly challenges. Before he was born, his father had moved the family west from Chicago in hopes of cashing in on California gold. When the prospect of riches dimmed, they sailed for Australian gold fields. That too failed and led the Smiths to Colorado, when Charles was thirteen. There he grew up surrounded by matters of geologic extraction. He spent cold Colorado winters as a machinist pressman. In the summers, he turned to mining and, by age twenty, grew more adept than his father, the failed gold miner, had ever been.

Mining was defined broadly in 1870s Colorado. Smith had worked under the engineering prodigy J. Alden Smith, a famed geologist and mining tycoon, during a boom time in the American West. The success of people like J. Alden Smith—no relation to Charles Smith and not to be confused with Senator William Alden Smith—seemed to suggest that all it took to find rich veins of gold, silver, coal, and quartz was a willingness to sit down and solve a few math problems. Charles saw miners who had no experience often get rich and famous. He watched as mining towns sprang up from nothing, just a few dirty tents on a ravine or near a creek that attracted grocers, tavern operators, and tool merchants, who eventually built brick-and-mortar stores along the strip. He had witnessed the ways in which wealth trickled down mountainsides as mining towns helped boost supplier cities like Denver, Boulder, and Colorado Springs. He saw major change come to the American West in the span of barely a decade and, with it, almost overnight prosperity. His attachment to the *Titanic* was thin. He was far more interested in solving one of the most vexing quandaries on earth and receiving credit for getting it done.

In late 1913, Smith spent three months drawing up more than fifty ideas. He imagined encasing the ship in ice to make it float to the surface. He pondered a giant claw that could be lowered and raised. He considered factors like deadweight, water pressure, buoyancy, and towing velocity. And he turned to the public for input. His goal wasn't simply to exhume the *Titanic* but to clean it, restore it, and return it to life again, as he told an interviewer:

> *My object, first of all, is to deliver the* Titanic *to its owners without further injury so that the great vessel may be rebuilt. Much of the cargo, or all of it, would be recovered. All the bodies which sank with the doomed ship have long since been embalmed by the action of the sea-water and when they are at last brought back to*

*the surface they will be easily identifiable and prepared for reverential burial.*

Smith approached this riddle the same way he approached a difficult mine. The first step was to find the desired material. Then he would figure how to extract it slowly and carefully, taking pains not to disrupt its integrity or damage it in a way that would devalue it. This made the *Titanic* different from a boulder veined with gold. You could smash a boulder into a million pieces and the gold would still be there. A fragile steamer ship wasn't as forgiving, but about this he didn't worry. He thought the final step—bringing it to the surface—would be the simplest. It might be slow and tedious, but all you had to do was counteract the pull of gravity. In mining, none of these operations were impossible; they simply required the right tools and a sensible plan. It helped that Smith maintained a surface-level approach to mining, a white-collar manager who stayed above ground, where he was insulated from subterranean discomfort.

Perhaps as a result, Smith's view of finding the ship was almost cartoonishly simple: Go to 41°46' N 50°14' W, the final coordinates the *Titanic* gave to the *Carpathia*, and activate a series of electromagnets attached by cables to surface ships. Once activated, they would be attracted to the steel hull of the sunken *Titanic* and, by force of attraction, slowly pull the surface boats to the exact site of the wreck. By all estimations, this would happen quickly, but there was a risk it might happen *too* fast, the magnets zipping through the water at high speed and capsizing the surface boats and dragging them across the water. Smith made several convenient assumptions that mitigated this possibility. The viscosity of increasingly dense water all but assured that the magnets would be slowed by immense friction, and as a result, that friction would provide the surface boats time to keep up.

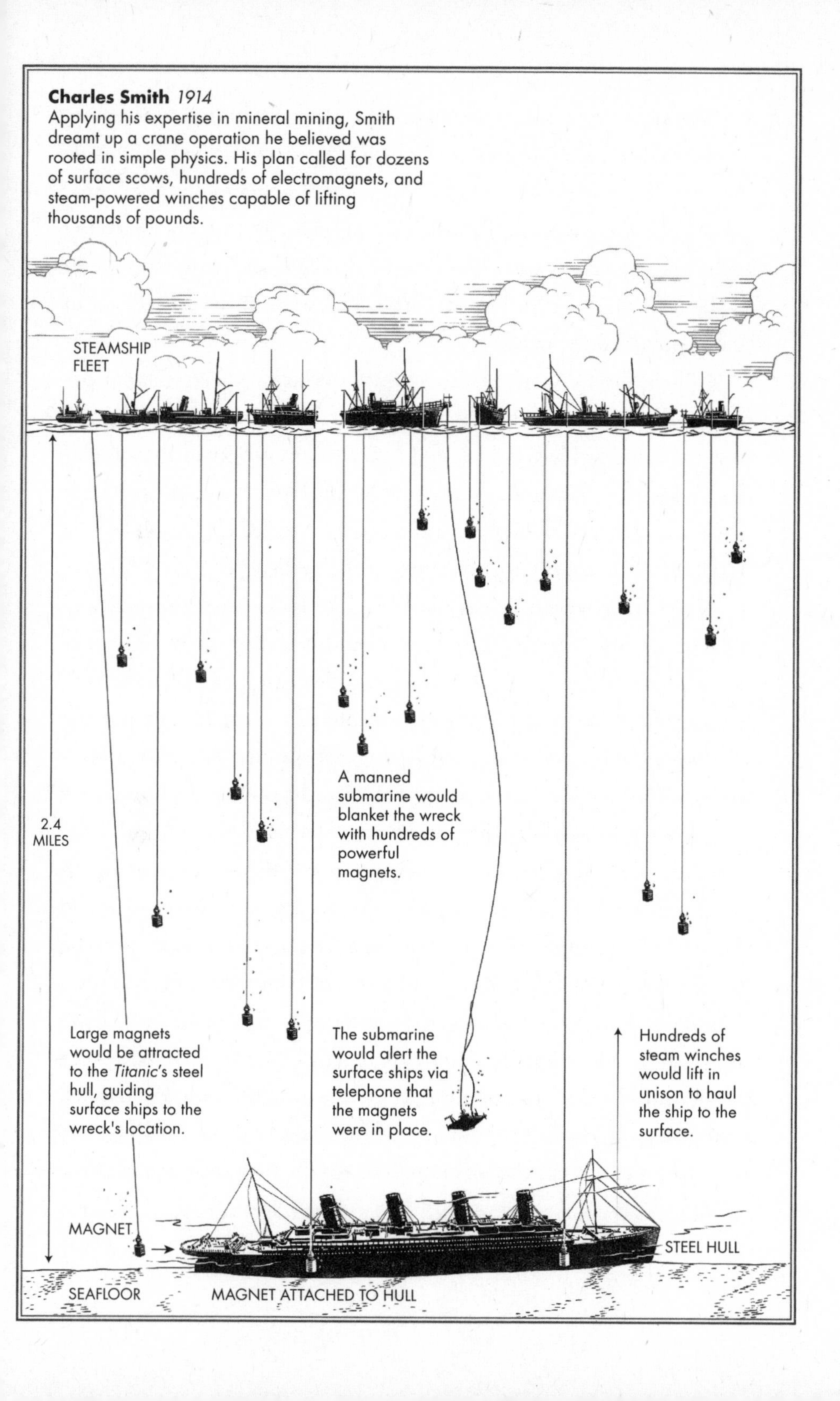
**Charles Smith** *1914*
Applying his expertise in mineral mining, Smith dreamt up a crane operation he believed was rooted in simple physics. His plan called for dozens of surface scows, hundreds of electromagnets, and steam-powered winches capable of lifting thousands of pounds.
STEAMSHIP FLEET
2.4 MILES
A manned submarine would blanket the wreck with hundreds of powerful magnets.
Large magnets would be attracted to the *Titanic*'s steel hull, guiding surface ships to the wreck's location.
The submarine would alert the surface ships via telephone that the magnets were in place.
Hundreds of steam winches would lift in unison to haul the ship to the surface.
MAGNET
STEEL HULL
SEAFLOOR
MAGNET ATTACHED TO HULL

The next phase was more elaborate. Once Smith found the ship, he planned to lower a tubular steel submarine connected by a cable that would enable communication by telephone and electric signal bells. Seven people could fit inside the submarine and, once arriving at the wreck, a mechanical arm would attach electromagnetic cables to the hull. Back on the surface, the cables would be connected to steam-powered winches.

"Then the real work begins," Smith declared in a lavish newspaper profile printed in newspapers in Little Rock, Pittsburgh, and parts of Kansas. He would spend exactly four weeks in the submarine, attaching hundreds of magnets until they "girdled" the ship. Then the surface scows would activate steam engines and the winches would slowly lift the ship off the seabed.

The lifting would be easy at first, but as the vessel approached the surface, Smith believed the ship would somehow grow heavier as the water grew less dense, an erroneous conclusion based on the era's incomplete understanding of deep-sea physics. In reality, the weight of a wreck stays the same, but lifting it becomes *easier* as it rises through less and less water. Smith believed the opposite, but he came to a novel solution nonetheless. The surface scows would raise the ship *partway* and then tow it in unison to a shallower area, at which point the winches would unspool and put the ship *back down* on the seabed. At this point, the surface scows would swap out the smaller cables for larger ones, add a few extra scows, and together, they'd lift the ship again, and perhaps a third and fourth time. Each lift would require a different arrangement of scows, and the sheer number of hours would require the whole operation to take breaks, laying the ship back down at night or for a lunch break. Finally, after surfacing the wreck, the scows would tow it to a shipyard in New

York, where it could be put in dry dock, scrubbed of barnacles, and restored.

When someone asked Smith how long the plan would take, he further simplified the timeline in hopes of attracting investors. Three months, he said. One month to find it. One to inspect it. One to lift it. "In three months after we start work I expect to have the *Titanic* on her way to New York in tow," he proclaimed from his Denver office.

Less than two months after he made his plan public in January 1914, more than twenty-five newspapers and engineering journals printed full details of the idea and a technical analysis of its prospects. They ran headlines like "Can the Lost '*Titanic*' Be Raised?" and "To Raise *Titanic* by Huge Magnets!," followed by lengthy discussions of the many technical challenges, but with better-than-not odds that Smith had the credentials to give the plan a high chance of success. "It is merely a matter of magnets," Smith said repeatedly to the line of newspaper reporters who visited his office seeking more details for their full-page stories.

A man of his era, Smith was rich in confidence. But when it came to actual money, he downplayed the price tag and emphasized the precision of his estimates as proof of his expertise. All it would take was $1.5 million and one hundred sixty-two men, he said. Smith didn't leave any room for mistakes, miscalculations, or overruns, because he didn't expect there to be any.

° ° °

Imaginative ways to conquer the sea and its contents were fashionable in Charles Smith's era. Engineers throughout history had seemingly solved much harder puzzles than this. The Egyptians had built towering pyramids, and the Romans giant temples and a Colosseum out of rock. The Incas put a settlement atop a mountain, and the Chinese bridged thousands of miles with a giant wall. Everywhere you looked in the early twentieth century there were signs of engineering triumphs. Railroads, printing presses, steam locomotives, mechanized reapers, telephones, airplanes, skyscrapers, and tremendous steamer ships. Smith's might not be the easiest operation, but surely engineers could return to the well of human ingenuity to figure out how to counteract a paltry fourteen million tons of downward force under a measly two miles of water.

This swollen confidence was especially strong when it came to ocean science. Benjamin Franklin studied the Gulf Stream current that traveled from Florida to the North Atlantic, fixating on the inconsistency that whaling ships that traveled in a rainbow-shaped route across the Atlantic arrived in Europe faster than mail ships that followed a straight line. A century later, in 1886, Prince Albert I of Monaco spent summers in the North Atlantic studying the variations in water temperature that he and countless scientists began to believe had something to do with global climate and, more specifically, weather.

The Gulf Stream was the object of scientific fascination throughout the nineteenth century, but by 1912, it had become so ho-hum that a mechanical engineer named Carroll Livingston Riker moved from studying it to wanting to manipulate it. On September 29 of that year, five months after the *Titanic* was felled, Riker released an "amazing plan," according to *The New York Times*. He proposed a two-hundred-mile-long rock jetty off the east coast of Newfoundland that would solve several problems at once. His feat of geoengineering would block the icy Labrador current and clear the way for the warm Gulf Stream to travel all the way north. This would melt icebergs before they floated into the paths of ships and, in doing so, would open new shipping lanes and ports across Canada, northern Europe, and Russia. Most exciting of all, clearing a path for the warm water would melt the ice caps and bathe America, Canada, and Europe in year-round tropical sunshine.

"I hear the exclamation 'visionary,'" he wrote, "but the idea is not visionary, on the contrary, it's exceedingly practical." At his urging, editors of *Scientific American* accompanied him on an expedition to Newfoundland to study the currents. They agreed that altering ocean currents would have dramatic effects on the weather, but they questioned the practicality of such a gargantuan project. Before the plan fizzled under the impossibility of sinking billions of pounds of sand and rocks in the open ocean, Riker declared the arrival of a new era of geoengineering. "Man Can Control All," he heralded in a widely circulated pamphlet. All that was needed was the will.

What's more, 1914, the same year Charles Smith presented his proposal, coincided with two major advances in ocean technology. One was the completion of the Panama Canal, a feat of engineering so Herculean it raised the sea to pass through mountains. The other was even more seismic: the formation by a Canadian inventor of sound navigation ranging, also known as sonar, which allowed a ship to map its surroundings by bouncing sound waves off nearby objects and correlating the timing of their return with the distance of hazards, including icebergs.

Such a string of ocean success explains why Charles Smith's ambitions were received not as glib and far-fetched but as bold and bound to succeed. It helped that Smith wasn't inventing any new technology. Magnets had existed for hundreds of years. Electricity and telephone lines could obviously work over hundreds of miles.

And the pressure? Smith had correctly calculated the deep-sea pressure at the *Titanic*'s two-mile-deep grave as somewhere in the range of four to five thousand pounds per square inch, which amounted to roughly fourteen million tons bearing down on the full hull. This would be challenging, he conceded, but only slightly. All he thought would be necessary was a slightly greater upward force—even just fourteen million and one pounds—and the ship would start to rise, probably quickly.

As proof, Smith pointed to an Italian inventor who, a decade earlier, had conducted almost the exact same operation Smith planned. His vessel, called the hydroscope, was shaped like a giant cigar, and, just like Smith's proposed submarine, it had electric lights and a mechanical arm. In January 1903, Giuseppe Pino asked the Italian navy to sink a small dinghy about one hundred fifty feet deep in the Gulf of Genoa. The next day, in front of a crowd, Pino boarded his craft and emerged an hour later with the dinghy clutched in the mechanical arm. A witness called it "like a fairy tale." Pino brushed himself off, lit a cigar, and proclaimed that a bigger hydroscope, perhaps enlarged tenfold, was all that now stood between man and the deepest ocean treasures.

At stake were wrecks like the *Madagascar*, a British frigate that disappeared on its way from Melbourne to London in 1853 with more than sixty thousand ounces of gold on board. And the *Brother Jonathan*, a paddle steamer that sank off the coast of California in 1865, taking two hundred twenty-five lives and as much as $7 million in gold to the bottom of the Pacific. And even the *Mary Celeste*, one of the great maritime mysteries of all time, found abandoned near the Azores in 1872 with nothing amiss except the complete disappearance of its crew. But none of those mysteries or lucrative opportunities matched the public fascination with rescuing the *Titanic*.

There were obvious differences between Pino's success and Smith's goal. The biggest was depth, a factor truly inconceivable in an era before anyone had gone to the deep sea. Breathable oxygen could be fed one hundred fifty feet underwater to the hydroscope through a surface line, but any deeper and the occupants would suffocate. There was also the small inconvenience of finding the ship—that the Gulf of Genoa could be combed in a day but the North Atlantic had no convenient boundaries. Had Smith taken the time to write to Pino for advice, the inventor likely would have responded with a hearty Italian belly laugh at the *stoltezza*, or foolishness, of the Colorado mining engineer.

Still, Smith had two secret weapons that would silence any doubt. The first was a media savviness uncommon among men of the early twentieth century. Smith knew how to pull the levers of publicity, particularly through newspapers, by dropping key words and themes that sparked public enthusiasm and curiosity. He had learned as a mine developer how even an offhand mention of a new vein of gold would bring to his door people willing to empty their wallets. And *Titanic* was an even more powerful word. Barely two years in the cultural swirl, any mention of the ship, its journey, its passengers, its crew, the food, the musicians, the iceberg, the watertight doors, the panicked telegraph call, the loading of the lifeboats, the screams, the fights, the names of the survivors, and the funerals of the dead all warranted front-page headlines. Nothing was too mundane or esoteric, from the final wardrobes of the high-profile passengers to the breakfast menu the day of the sinking to a lengthy dissection of the metal quality in various parts of the ship. One day, Smith was delighted to realize that a basic napkin sketch of his plan to retrieve the *Titanic* that he had shared with a friend yielded a visit from a reporter from the Colorado paper *The Ordway New Era*. Smith then clipped the resulting news article and posted copies to several big-city newspapers and then watched the news of his "brilliant" and "ambitious" plan trickle down into working-class news rags like the *Plano Star-Courier* in Texas, the *Stevens Point Daily* in Wisconsin, and the *Abbeville Progress* in Louisiana.

Publicity could get a man far. But even a million newspaper articles couldn't get the *Titanic* off the floor of the Atlantic. Smith second's weapon was his flair of confidence. Smith had learned from experience that the endless loop of promotion and excitement could be a substitute for the intricacies of complex engineering. This confidence alone moved his ideas from outlandish fantasy to technical reality and, from there, to doubtless certainty.

○ ○ ○

Despite Smith's flamboyance, his sense to raise a ship, even a large one, was not naïve. In the right context and for the right price, a sunken vessel could be made to float again. The technology to salvage underwater vessels was developed in the late nineteenth century and was deployed most often to investigate accidents or recover lost goods. Even amid the industrial revolution, most ships were sent to sea dangerously overloaded or with underprepared crew. Owing to this, U.S. companies had grown accustomed to losing more than 6 percent of all overseas exports to shipwrecks. The motivation to enhance salvage capability came not from oceangoers but from marine insurance companies that had grown weary of paying high claims for lost cargo.

America's first school for underwater divers was established in Newport, Rhode Island, in 1882, a town with abundant access to calm and shallow water. The divers relied on primitive tanks of oxygen and tethers of surface air to survey wreck sites. They were trained to recover unspent torpedoes and, for the most advanced among them, salvage entire ships. This went surprisingly well. By 1907, the divers attached a series of cables to the *Machias PG-5*, a gunboat that had sunk a year earlier in a hurricane near Pensacola, Florida. The ship was refloated, pumped, patched, and put into operation for twenty-five more years.

Operations like these were expensive. But so were ships, many costing hundreds of thousands of dollars to build and maintain, which rationalized the effort to give them second and sometimes third lives. After the battle of Manila Bay in 1898, during which Spain scuttled three gunboats, the *Isla de Luzón*, the *Isla de Cuba*, and the *Don Juan D'Austria*, the victorious U.S. Navy salvaged the ships, restored them, and, in a flourish of sovereign superiority, kept Spain's original names but with an all-important prefix, United States ship, or USS.

Not all operations were successful. The navy was poorly equipped to navigate deep-water wrecks. Anything below three hundred feet was considered unreachable, a benchmark set when the submarine *F-4* sank near Honolulu on March 26, 1915. Divers pulled cables more than two hundred feet underwater with a plan to physically drag the boat to higher ground, where remains of the crew could be removed and its hull repaired. But when the cables snapped and the sub remained stuck, navy engineers concluded that the vessel was filled with water and could only be moved with pontoons.

The thorniest salvage operations can contort the meanings of failure and success. In one of the most eye-popping salvage sagas in history, the USS *Dewey*, a floating dry dock capable of repairing damaged ships in open water, sank near its port in Manila in 1909 after being accidentally struck by a torpedo. Navy seamen tried twice to refloat the craft by pumping air inside. They succeeded on a third attempt and put the *Dewey* back into operation. Years later, after the vessel had weathered World War I, the *Dewey* fell to the seabed again in 1942 when its crew scuttled the vessel to prevent it from falling into Japanese hands. Then things got strange. After sinking for a second time, the *Dewey* was refloated *again* in 1944, this time by the Japanese, who thought the American vessel could be useful in the escalating naval war. The *Dewey*'s final ironic insult came from an American bomber that fired a torpedo at the long-serving American vessel and sent it, for the third and mercifully final time, to the bottom of the Pacific.

Yet for every case of salvage motivated by money, power, or wartime strategy, there are ones driven purely by emotional catharsis.

Most Americans had never heard of the armored cruiser the USS *Maine* before it sank in a dramatic explosion in Havana Harbor in February 1898. The incident, thought to have been caused by a Spanish torpedo (though historians have more recently blamed a munitions accident aboard the ship), made the *Maine* an instant hagiographic symbol of American imperial outrage. The offense sparked a summer-long battle against Spain, and after America prevailed and signed a treaty that awarded them Spain's colonies in the Philippines, Puerto Rico, and Cuba, Congress started to wonder if it was worth getting the *Maine* back.

Pulling it from its underwater grave wasn't just an exercise to salvage some wood and steel or even old weapons. The two hundred sixty sailors who died on the *Maine* couldn't be saved, nor could a once-functional ship that had been damaged by an explosion be returned to operation. But by salvaging the wreckage, the story of a ship could be extended. Like a soldier who dies on the battlefield, the corpse of the *Maine* could be collected by its comrades, carried somewhere honorable, and laid to rest in peace.

Removal proved possible, but revival was not. In a major but largely forgotten engineering triumph of the twentieth century, the U.S. and Cuban governments combined forces in 1910 to refloat the *Maine*.

Had the ship been in better condition, the easiest approach would have been to sink large water-filled structures called caissons, attach them to the ship with cables, and slowly pump the water out of the caissons to make them—and the ship—float to the surface. But this would happen too fast and pull the wreck apart. Preserving the wreck required it to be lifted far slower. This called for a so-called cofferdam, an ancient Roman technology of building a barrier in a body of water and draining it from the inside. For the *Maine*, sunk in thirty-five feet of water, this entailed building forty-foot walls around the perimeter of the ship in a rounded shape to share the weight of the water pushing it from the outside. Once the dam was built, the inside of the walls would be drained to expose the ship sitting on the mud. And once it was dry, investigators could inspect the wreck, bodies and equipment could be recovered, and the hull could be repaired and prepared to be refloated.

This would be one of the largest engineering projects the U.S. ever embarked on in a foreign country, let alone underwater. Naval engineers brought five thousand tons of steel to Cuba from Buffalo, New York, via the Saint Lawrence Seaway, a journey of more than four thousand miles. The parts included seventy-five-foot steel piles that were fashioned into large cylinders and, upon their arrival in Cuba, were filled with clay from a nearby dredging barge. Next to the wreck of the *Maine*, the seventy-five-foot piles were planted so deep in the mud that they poked out of the ocean surface only eighteen inches. Then they were braced by extra rocks at their base and wood beams. Cofferdams had been built before, but never this big and never in such soft mud and clay. Even on June 3, 1911, the day the pumps started "dewatering" the inside, engineers were only partially confident it would hold.

It held. After weeks of slow pumping and adding wooden planks to brace, on August 5, when the cofferdam was nearly dry, investigators bravely descended to inspect the ship. They recovered weapons and the remains of several dozen crewmen, virtually all of them trapped as the ship sank. The *Maine* was effectively embalmed and laid to view in an open casket as Americans gawked at photos in their newspapers of the *Maine*'s autopsy. For officers inside the cofferdam, it was easy to notice the corrosive effects of the ocean, even just a few feet deep. All wood and steel lodged in the mud was in the same condition as the day it sank. But the portions above the mud were visibly eroded by wood-boring shipworms, rust, salt, and sun.

After months of investigating the ship and taking photos and notes that would be scrutinized by munitions experts and a century of historians, the *Maine* was prepared for its final voyage. Engineers repaired damaged sections of the hull with steel plates and removed other sections too broken to fix. Finally, on February 3, 1912, after all valuables were removed and bodies recovered, water was slowly added to the cofferdam to refloat the *Maine*.

To avoid any sudden jerky motion, engineers refilled the cofferdam at the same trickle as filling a bathtub. It took a full month as the ship inched upward, until eventually the *Maine* sat on the water in exactly the same spot where it had exploded more than a decade prior. A navy tugboat called *Osceola* started pulling the rebirthed wreck. Two other tugs attached cables to the sides to keep the old boat steady. As it left the harbor adorned with an American flag and bouquets of flowers, American crewmen stationed in Cuba lined the rails and saluted. A band played the national anthem.

Two hours later, the *Maine* was three miles offshore, a sufficient distance to ensure that, once sunk again, it wouldn't bother anyone. A wrecking crew boarded the ship and began opening the valves and sluice doors to flood the ship. They departed quickly on waiting skiffs and then there was nothing left to do but watch the wayward structure bob and wallow. Several other warships joined the scene, and their entire crews watched the last send-off of an American symbol. It was silent except for the roll of the sea.

In the same year in the same season, separated by barely four weeks, the *Maine* sank for the second time, on March 16, 1912, in almost the exact same way as the *Titanic*, one ship beginning its cultural voyage and the other one ending. The *Maine* tilted as water filled its bow. The stern rose in the air until it was almost vertical, showing its propellers and then its keel. The American flag flying from its mast dipped into the water. Moments later, the stern disappeared, and the ship began its 3,770-foot fall to the floor of the Atlantic. There was nothing more to be said and nothing more to see, except for the blanket of flowers where it had once been, floating on the surface of the water.

Chapter 5

# LUNGS THE SIZE OF ACORNS

Judging from the letters and telegrams that appeared in his office each morning, Charles Smith felt satisfied that news of his calculations had reached far and wide, appearing in papers as far flung as Europe and Asia. Even more impressive, people had lined up to give him money. Smith assured investors that after the ship was raised, towed, and revived, the White Star Line or whoever took over operation of the unsunken liner would owe credit and perhaps a handsome reward to the person who rescued it.

This made Smith a popular horse to bet on. Wagering had grown into an American industry in itself at a time when the gilded wealth of America in reconstruction had evolved into a booming economy with risk-takers and groundbreakers. American aviation had grown from a backyard hobby in 1906 to a flourishing industry in less than a decade based on heavy investment and fast advances to formerly unthinkable problems. American companies were rising at a pace never before seen, with venerable names—Hallmark, Black & Decker, Kellogg, Neiman Marcus—that would dominate for a century, all underscoring that boldness, however brazen, was rewarded.

Of the $1.5 million Smith needed, he raised a little over $10,000 in the span of six months. The sum seemed large at a time when the average American salary was barely $700 a year, except for two wrinkles. One, almost a third came from Smith himself, who offered to finance $300,000 of the project. Much of the rest came from Smith's mining partners and neighbors. People sent letters along with a quarter or two, asking him to add their names to the list of investors, hoping to double or triple their money—or better, to own a piece of the famous *Titanic.* Smith at first accepted these missives, but after a few dozen, he found it to be an inefficient way to raise a million dollars and a poor use of time. He marked the donations *return to sender.*

In all of Smith's technical papers from the time, one detail that seems to be missing is exactly what role Smith saw for himself. He was the ringleader, goal-setter, and fundraiser. He was a timeless character of a hype man whose success was the result of braggadocio and blame-shifting. But even his swollen confidence couldn't obscure the fact that he had no ocean experience, had never gone underwater in a submarine, had never directed an armada of boats bobbing on the surface, and had never salvaged even a canoe. He was, by all appearances, out of his depth, and even though he had decades of experience leading large projects and making vast sums of money, he was self-aware enough to know he wasn't a workhorse. He saw himself as a theorist and ideas man, and when it was time to get to work, he would just as well give the details to someone else to get it done.

This would become common among fans of the *Titanic*, who in the 1910s weren't so much enthusiasts about ocean science and marine dynamics as they were deeply drawn to the disaster and its ripples of sorrow and destruction. They were pulled in by the human elements of wealth and loss, which passengers died and which lived, which men disgraced themselves by joining lifeboats, and how others behaved with valor in their final moments.

One group entirely unmentioned in the public requiem for the *Titanic* was professional salvage companies. By 1915, there had not been a serious industry effort to rescue the ship and bring it to the surface. People who knew anything about marine physics knew that it wasn't even worth trying. What's more, the most casual student of ocean science might have pointed out the string of Smith's fantastical assumptions, including that the *Titanic* was still in one piece and could be found quickly, or that a yet-to-be-built submarine could descend to crushing depths, or that electromagnets could lift the forty-six-thousand-ton wreck, or that towing would be easy and smooth, or that caissons with the amassed buoyancy of 215 *Titanic*s could be made and transported to the site, or, most of all, that the money and man power for the operation would come easy and fast.

Contrary to his expectations, Smith's critics did not center their mockery on his credentials or the awkward geography that a man claiming omniscience over the ocean lived in a landlocked state. They took the idea at face value and dismantled it piece by piece. The editors of *Scientific American* called it a "ridiculous proposal" and pointed out that electromagnets were known to emit force over a few hundred feet, not miles. Finding a ship the way Smith envisioned would be like someone holding a magnet in the middle of a public park hoping it would stick to the nearest refrigerator.

Simon Lake, a mechanical engineer and maritime architect who built the first wooden submarines for the U.S. Navy, questioned whether the *Titanic* was findable at all, and if it was, he thought it would take Smith significantly longer than a month. "It's possible that the Gulf Stream and the Labrador current have in the past two years substantially buried the wreck of the *Titanic*," Lake wrote. "If such be the case, the hulk may never be located." Lake had once tried to find a wrecked cargo ship that had sunk in 1909 in Long Island Sound. Other hunters spent two years diving in fifty-foot water to find the craft and salvage its rumored cargo of copper and iron ore. When they turned up nothing, Lake deployed two wreck-finding submarines he had designed out of pinewood. Wielding such advanced machines, he figured the process would take him hours, not days. Several months later, he expanded the search area and finally found the wreck fifteen miles away and more than twice as deep as the site where it was said to have sunk. He salvaged the cargo, but raising the hull was impossible. Lake published his findings in 1914, the same year Smith was shopping his plan, and declared that water currents carry wrecked ships along the ocean floor. Lake thought the *Titanic*'s search area was likely hundreds of miles wider than Smith had figured.

If it *could* be located, there was the question of the hardness of the seafloor. If the ocean bottom was covered in mud, the ship may have landed softly, like a bowl on a pillow. If it was sandy, the ship falling bow first may have pierced itself into the seabed, like a stake in the ground. If it was rocky, the hull may have broken in a million pieces on impact, leaving no structure for magnets to grab on to.

The topography of the seafloor was a long-standing mystery. The common assumption was that it was flat, a uniform bathtub with no mountains, valleys, or other geographic features. In *The Origin of Species*, Charles Darwin's theory that diversity comes from reproductive isolation applied to terrestrial species only. By comparison, Darwin seemed to believe that the oceans were homogenous, like a giant fish tank with no discernible boundaries. If a fish wanted, it could reasonably swim from Japan to Portugal, which implied similar ecosystems and species in both places. If every marine environment was similar, then no environment was unique, and thus, the composition of the seafloor could be dismissed as irrelevant.

Lake and other researchers also questioned the scale of Smith's calculations. The most advanced engineering firm in England had demonstrated that lifting magnets could be used to recover scrap steel and iron lost overboard in harbors, but the magnets were enormous—more than five feet in diameter and weighing twelve tons apiece. To lift something as large as the *Titanic* would conservatively require 3,250 magnets that collectively weighed ten thousand tons, weight that once dispatched to the seafloor would have to be hauled back up.

Try to find enough boats to do that, Lake taunted Smith in a lengthy take-down in the March 15, 1914, edition of the *Buffalo Sunday Morning News*. He'd have to employ every single boat in New York, Boston, and Annapolis combined. Then he'd need to coordinate them into the largest flotilla in history to the middle of the Atlantic. Others piled on in the *Nashville Journal*: "If you are at all familiar with the practical difficulties of towing you will realize what it would mean to convoy these squadrons to the site of the wreck, and when you have crowded them over the *Titanic*'s hulk, imagine what would happen if a storm arose?"

The challenges were abundant but they weren't insurmountable, the papers concluded. It was possible that enough boats existed, enough magnets could be found, enough crewmen could be recruited, and enough money could be raised. But there was one factor that could not be quelled by the most thoughtful plans of the most advanced scientific minds. The monster that dwelled at the bottom of the sea would threaten to diminish the effort from ambitious to preposterous, from expensive to priceless, and from difficult to impossible, and the monster was pressure.

∘ ∘ ∘

Anyone could calculate the force of pressure underwater in 1914, or even in 1814. Pressure works on an arithmetic scale, increasing incrementally with depth. At sea level, the weight of all the air, clouds, and moisture in the atmosphere presses down to exert 14.7 pounds per square inch (psi) on every person, dinosaur, and woolly mammoth that has ever lived on earth. Below water the pressure grows by another 14.7 psi every ten meters. Add, divide, multiply, and the *Titanic*, twelve thousand five hundred feet deep, sits under the combined weight of nearly six thousand pounds per square inch, or roughly the weight of a fully loaded SUV balancing on your pinky finger.

That quantity of force is enough to kill a person, but not by pancaking them flat as you'd think. Globular pressure underwater applies equal pressure on all sides, which wouldn't hurt the aforementioned pinky finger were it not for the tiny pockets of air inside. Without the inconvenience of gaseous oxygen, nitrogen, and carbon dioxide filling our lungs and coursing through our veins, people could reasonably descend miles underwater. But these gases compress forcefully. On the surface, an adult pair of lungs are the size of two footballs. A hundred feet down they shrink to the size of two baseballs. Two hundred feet lower they're barely the size of acorns. Descend twelve thousand more feet and the lungs would be airless masses of congealed tissue. The cause of death wouldn't be a collapsed lung but a failure to breathe. At the same time, your stomach, intestines, sinuses, eardrums, and every other air-filled organ would be squeezed unworkably shut.

In 1914, this was an inconceivable quantity of force. One could feel six thousand pounds per square inch in only a few ways, like being trampled by a herd of elephants or getting struck by a locomotive. Few who experienced such trauma lived to recall it, and if they did, the moment of impact was shrouded in blackout. The notion of sending a person down in even the most rigid steel suit was as far-fetched as sending a man to an asteroid. Theoretically it was possible, but no one had the capacity to do it.

In the following century, mankind would make considerable strides on both creating higher-pressure environments and insulating the human body from their crush. In the 1920s, pistol makers began advertising guns with chambers exploding under an incredible twenty thousand pounds per square inch. Not long after, a Wisconsin company introduced a high-pressure water saw capable of cutting through metal and granite with the force of sixty-five thousand pounds per square inch. Around the Second World War, an air force doctor named Harry Armstrong accidentally discovered the profoundly strange ways pressure is influenced by altitude. When Armstrong got suddenly hot, passed out, and almost died in a high-flying airplane, he realized that at sixty-three thousand feet—a number subsequently known as the "Armstrong Limit"—the low atmospheric pressure effectively reduced the boiling point of water to the temperature of the human body, putting a pilot at risk of his blood boiling. His discovery that low pressure can badly damage a human body led to airlines pressurizing their cabins, future astronauts wearing pressurized suits, and climbers of Mount Everest requiring advanced breathing systems to summit the peak.

But even these are small numbers. Scientists have since figured out how to make artificial pressures as high as one hundred million psi by pressing together the tips of diamonds. Their field, known as "extreme state matter," searches for ways to craft better solar panels, building materials, and the perennial holy grail of nuclear fusion. Astoundingly, even *higher* pressures exist naturally in the universe: Jupiter's core is around a billion pounds per square inch, which is small compared to a neutron star, the collapsed matter of a burned-out sun, whose center holds a billion trillion times more pressure than Jupiter's core. These numbers are impossible to fathom and entirely irrelevant. The most colorful way to appreciate pressure is to observe, as General Electric demonstrated in the 1950s, that under high enough compression, peanut butter turns to diamonds.

Charles Smith wasn't battling a neutron star. By comparison, six thousand pounds per square inch is paltry, almost marginal. But it remained impossibly big. Whales could withstand dramatic pressure thanks to their flexible ribs. Squid could swim freely at great depths without air sacs. But humans could not. Of the few hundred people who considered themselves professional divers in the nineteenth and early twentieth centuries, only a few dozen had gone down one hundred feet, and barely a handful had gone past that. Some died trying. In 1901, a youth magazine described the horrors of diving in cheeky terms: "The wife of a diver, poor woman! [She] starts with terror every time she hears a doorbell ring."

There was no justifiable reason to risk life and limb except to see something interesting, like sunken ships. In 1891, the American journalist Cleveland Moffett documented his attempt to attach chains around a sunken tugboat near Fort Montgomery on the Hudson River in New York. The first diver who tried had passed out at one hundred feet and had to be pulled up by a rope tied around his waist. ("I felt like I was dreaming," the man said when he came to.) Moffett then volunteered to go down, put on the one-hundred-seventy-five-pound metallic suit, and descended to one hundred fifty feet.

> *I got into the suit and went down, and I stayed down until that chain was under the shaft. It took me twenty minutes, and I don't believe I could have stood it much longer. The pressure was terrible, and those twenty minutes took more out of me than four hours would, say at fifty feet. But we got the tugboat up and she's running yet.*

Smith didn't intend for a diver to go down to the *Titanic.* But nor was he able to deliver a craft capable of the underwater maneuvers he had in mind. By 1914, virtually all the world powers had submarines in their navies that labored to go past seventy-five feet. Even if someone could build a steel cylinder capable of withstanding deep-sea pressure at one or two miles, there was the question of buoyancy and how it would get back up. Submarines function with intricate configurations of air ballast; flooding the tank makes it sink, air makes it rise. Provided Smith could find the right balance to descend slowly, he would have to adjust the air tanks manually to make the craft hover slightly above the wreck.

Smith had capitalized on the human longing to fill an open wound. He fanned the insatiable hunger to hold on to the ship, up to and including physically grasping it again. But provided it could even be done—safely, efficiently, affordably—the questions remained as to what good any of it would do. An editorial referencing Smith in the *Escanaba Morning Press* of northern Michigan published a sniping one-line editorial comment: "A western genius declares he can raise *Titanic*, but he neglects to tell us why anyone should want to raise her."

Nostalgia is a powerful force, yet it's still no match for the forces of the deep sea, to say nothing of the realistic price tag. The *Iola Register* in southeast Kansas posed it the most directly: "Personally we have no inclination toward heroism, and if we could raise the money to raise the *Titanic* we'd keep the money and let her stay there." Smith had gone to great lengths to map the intricacies of the who, what, where, and when. Yet in trying to muster the case for the world's most ambitious civil engineering project in the history of humanity, he couldn't explain the why.

∘ ∘ ∘

More than wistfulness, nostalgia, financial prudence, or protection of secrets, the most common reason to salvage and refloat a ship is because it blocks traffic. That was why the Cuban government wanted the *Maine* moved from Havana waters, and also why obscure ships like the USS *Saint Paul*, of no particular distinction or renown, received the resources necessary to remove, refloat, and revitalize it after it capsized in its dock in New York Harbor in April 1918.

Shallow water helps. Neither the *Maine* nor the *Saint Paul* would have been raised and refloated for duty had they sat hundreds of feet deep. Accessing them would be too expensive, and they would not obstruct ships still in operation.

To figure how deep is too deep to recover, one need only consider the largest ship ever built, a Japanese-made supertanker the length of four football fields called the *Seawise Giant* (and later renamed *Knock Nevis*), which floated from 1979 until it was sold for scrap in

2009. The *Titanic* had once been the largest moving object on earth. By tonnage, the *Seawise Giant* was six times bigger, with a turning radius of two miles. When afloat, its hull descended eighty feet underwater, which implied that so long as any wreck was deeper than that, no ship on earth could strike it.

There is a gray zone, however, of what constitutes an obstruction to surface navigation. Sometimes the obstacle isn't so much physical as emotional. The Swedish government faced such a dilemma after the sinking of a steamer ferry, the SS *Per Brahe*. The ferry was so overloaded with crates of stoves, sewing machines, peaches, and pears that on its way to Stockholm one night in November 1918, a gust of wind blew over some of the cargo, rocked the ship off-balance, and sent the capsized vessel and its twenty-four passengers to the bottom of Lake Vättern. The incident would have been unremarkable had one of the dead passengers not been Sweden's most famous artist at the time, John Bauer, who was moving to Stockholm with his family in hopes of a spiritual revival of his work. Owing almost entirely to Bauer, the *Per Brahe* grew into a *Titanic*-like fable of hubris and humility, which compelled Swedish officials to find a way to get it back. But the more compelling reason the prospect of raising it drew government support was that it was moored in one hundred five feet of water, a depth dangerous to other ships not physically but preternaturally, considering the wreck was a grave site. Plus, unlike the *Maine*, the *Per Brahe* had sunk not from an explosion but from its own obtuseness and was believed to be in revivable shape.

The day the *Per Brahe* was raised in 1922, twenty thousand people lined the shoreline to watch it be towed into port. To finance its salvage, newsreels of the second coming of the now-famous ship were shown in cinemas across Sweden. Much like the *Titanic*, endless fascination with the *Per Brahe*'s demise fueled what one newspaper called a "macabre tour," where the ghost ship was towed from port to port for people to gawk at. And when interest subsided and spectators stopped buying tickets, the *Per Brahe* was repaired, rejuvenated, sold to Finland, and quietly put back into service as a cargo steamer. The hope was that the ship could shed its chilling past and just get back to work.

The *Titanic* was never a serious candidate for such an operation, and with its financial constraints, Charles Smith never had a chance. The $1.5 million Smith sought amounted to a quarter of all U.S. spending on health care and a fifth of the country's entire budget for education. If the technology existed—and Smith's public dressing-down in the media had exposed that it did not—the actual price tag of finding, raising, and restoring the *Titanic* in 1914 was probably in the range of $50 to $100 million, not counting inevitable overruns and delays. Such a sum was beyond any one man's ability to underwrite, and the U.S. Treasury could not be hijacked for such a complex project highly prone to failure.

One irony of Smith's collapse was that his timing was, by the measures of a sunken ship, as good as it would ever be. There was still frequent debate among eyewitnesses and ship designers about how the *Titanic* actually went down. It was difficult to imagine how it might have broken in half, particularly when the iceberg scraped the bow and not the center point, where it might have weakened its central hull. To the average reader of scientific journals and newspapers, it was easier to picture the ship sinking in one piece, as most ships do, and that it remained intact on the ocean floor.

Had this been true, in the hypothetical alternate reality that the *Titanic* had remained on the seabed in largely the same condition it was seen leaving Southampton, the best possible time to salvage it would have been in the immediate years after it sank. Erosion of the wooden decks, the steel railing, and the iron hull would have begun instantly, but years of erosion are favorable to decades. Later in the century, when underwater technology advanced far enough to support such a sweeping search-and-salvage mission, the aging ship's devolution and dereliction would leave it too diminished to be moved.

Charles Smith might have realized that of all the people who would ever dream of lifting the *Titanic*, he probably had the best shot. Even a mining engineer in Colorado had to know that leaving an object deeply submerged in seawater for decades would eat away at its structure, the same way an abandoned mine would eventually collapse and reclaim itself. The planet has a way of humbling the bold, and before long, Smith found himself publicly humbled. Every component he promised would be simple and cheap turned out to be unimaginably complex and exorbitantly expensive.

By late 1914, Smith's once-grand plan had been reduced to small, single-line snippets in the newspapers with vague updates—often reporting there were no updates. Several months later, in the early days of 1915, all mention of Smith and his plan had disappeared altogether. The full spreads and elaborate illustrations of electromagnets and electric cables were taken over by the run-up to the biggest foreign war America had yet experienced. On February 22, 1915, after the Germans spent months placing floating mines in common shipping routes, Americans woke to news that two U.S. cargo steamers, the *Carib* and the *Evelyn*, had been struck and sank. Within two weeks, President Wilson authorized an enormous expansion of the U.S. Navy, including the rapid construction of two new battleships, six destroyers, and eighteen submarines. Two weeks after that, the assistant secretary of the navy—a rising young politician named Franklin Roosevelt—began to publicly lobby Congress for more American muscle at sea.

Three years had gone by since the *Titanic* sank, but it may have been three hundred considering how quickly the world had changed. Shock that a steamer had sunk in 1912 was overtaken by almost daily sinkings in 1915. The battles between American and British ships against German U-boats yielded incessant tragedy. In late March, the British passenger ship the *Flaba* was sunk by a U-boat (*UB-28*) near Ireland. In early April, the German auxiliary cruiser SMS *Kronprinz Wilhelm* stopped near Virginia after eight months at sea and was immediately boarded and overtaken by the U.S. government. Fighting would continue throughout the summer and the following three years. Most were small maneuvers compared to the tragedy of May 1915, when a U-boat torpedoed the British passenger steamer *Lusitania* off Ireland. Two explosions sank the ship in eighteen minutes, killing almost twelve hundred civilians.

The *Lusitania*'s dead rivaled the number lost on the *Titanic*, but the *Lusitania* seemed worse. An iceberg was an act of God, but a torpedo was at the hands of man, and thus avoidable. Early reporting in England and the U.S.—whose loss accounted for 128 of the total—decried that the *Lusitania* was a civilian vessel sunk in an unprovoked attack in violation of wartime custom and international law. But the German government argued that the ship had entered a declared war zone, and, in addition to its full load of passengers, was also alleged to be carrying hundreds of tons of munitions, which made it subject to attack. The British insisted on the more favorable storyline for sixty-seven years, until 1982, when salvagers were getting close to excavating large portions of the *Lusitania*'s wreck. For their safety and to avoid another tragedy from the same wreck, the British foreign secretary admitted that while it was true the *Lusitania* had no weapons, it *was* transporting more than seven hundred tons of gun ammunition, 1,250 cases of shrapnel artillery shells, and more than one hundred cases of explosive bronze and aluminum powder.

It's hard to imagine someone like Charles Smith ginning up enthusiasm to raise the *Lusitania* the way he did the *Titanic.* Technically, it would have been easier, sunk in a mere three hundred feet of water. But sitting barely twelve miles off the coast of Ireland put it tantalizingly out of reach for a U.S. engineer. And besides, the *Lusitania*'s demise came in a long string of tragedies with wartime deaths in the millions. If there was money and time to be spared, it would be to fortify and build new ships, not rescue dead ones.

This explains why Charles Smith disappeared from public view less than a year after he proposed his ambitious plan. He was nearing sixty and the final third of his life. Mining engineers don't often find fame as enchanters of the world's imagination, but Smith had enjoyed his time in the sun, and when it set, he knew it was over. Rather than double his efforts to capture the nation's fraying attention, he let go and walked away. He returned the remaining money to investors and closed his accounts. Smith had the benefit of never being especially obsessive about the *Titanic*—this persona would come later, and with gusto—as much as he was drawn to a great challenge of engineering. But when the project went south, he did the same thing he did with a dried-up mine: he packed up and moved on. The price of zinc doubled in 1915, and Colorado had some of America's largest deposits. Smith returned to work and let go of a plan that seemed a frivolous relic of a former time.

As Smith yielded the stage, it was quickly taken over by bigger ambitions in the deep sea, ambitions that had nothing to do with nostalgia. In April 1916, a band of twenty-four self-described "young capitalists" on Wall Street, led by Percy Rockefeller, the nephew of John D. Rockefeller, formed a corporation called the Interocean

Submarine Engineering Company to find the wealth lost in the sea. Already four hundred eighty-one vessels had sunk as casualties of the Great War, their collective value more than $243 million. One of them, the passenger steamer SS *Merida*, which shuttled between New York and Havana, was thought to be carrying 372 bars of silver worth more than $200,000 and more than $500,000 in other coins and other metal bars.

Deep-sea treasure hunting was a lust for money wrapped in a self-serving vision of technological conquest and a dose of national philanthropy. Like Smith, Rockefeller and the Interocean Submarine Engineering Company earned considerable press at a time when dreaming big was in vogue and overpromising bore little risk. And like Smith, the company ran into the head winds of world war, insufficient funding, overruns, and embarrassing miscalculations.

As a decade, the 1910s were heavy with more nautical defeat than triumph. The worst of them came in the midst of the biggest war the world had ever seen. In 1916, an Italian troopship, the *Principe Umberto*, was passing from Albania to Italy when it was struck by an Austrian torpedo. Lives were lives, but military casualties seemed not to warrant the same public outrage as civilian lives, especially wealthy ones. In a sober example of which shipwreck victims get remembered and which are forgotten, the *Umberto* took 1,926 soldiers' lives to the bottom of the Adriatic—more than four hundred more dead than on the *Titanic*. The next day, *The New York Times* reported the incident in four sentences on page ten.

Chapter 6

# I REGARD THE *TITANIC* AS MINE

Lawrence Beesley didn't want to die. All he wanted was to be part of the action, which at the moment was on the starboard deck of the RMS *Asturias*, a retired half-scrapped troopship that with a fresh coat of paint and new lettering was now an identical replica of the *Titanic*. Beesley was caught up in the excitement, the sense of panic and mayhem. He recognized that the next few minutes would memorialize the seminal moment of the twentieth century, and with him at the center of it.

Decades earlier, Beesley had bought a ticket to tour the United States. He might have crossed the ocean aboard the *Olympic*, the previously grandest ship of the Atlantic, but friends convinced him to wait for the next-best model. Beesley, from Derbyshire in the East Midlands, had been raised in a family wealthy enough to make such lavish choices. When it came time to sail aboard the *Titanic*, he arrived hours before departure to revel in the trappings of opulence, wandering the dining salons and libraries and the gymnasium, where he was photographed on the electric camel that would pulse a person forward and back as a form of exercise.

Days later, as the ship began its slow but sure descent to the bottom of the sea, Beesley was surprised by a turn of luck. Rather than go down with the ship, as was expected of a man of thirty-four yielding to younger and female passengers, he happened to be standing near lifeboat thirteen, which he noticed still had several seats available. Not a lady or child was in sight, so Beesley did as instructed and got in the damn boat. To survive was better than the alternative, but it resigned him to the lifelong fate of retelling the details of this story as a means of reputational self-defense.

Which is why, when seeing a *second* chance to sink with the *Titanic*, he was ready to take it.

In January 1958, Beesley arrived at Pinewood Studios west of London dressed in khaki pants, a tie, and a baggy black peacoat down to his ankles. He was eighty, with silvery slicked-back hair, and he was prepared to meet his fate, or some semblance of it, and finally perish with the ship.

To film *A Night to Remember*, this second sinking of the *Titanic*, was the culmination of five months of re-creating every detail of 1912 to the point that the catering company provided the crew with the exact same meals served on the ship. Now, in 1958, the sinking scene would forever be the most historically precise reenactment of the real thing, down to the exact words spoken in the ship's final moments as they'd been recalled by survivors. No movie studio in England had a pool big enough to film passengers clamoring for lifeboats, so the production was moved to Ruislip Lido, a nearby reservoir, and scheduled for two a.m. on an icy night. This required hundreds of background actors to jump into the frosty water. When they hesitated,

Kenneth More, the main star of the film, who played second officer Charles Lightoller, jumped and instantly regretted it. "Never have I experienced such cold in all my life," he later said. "It was like jumping into a deep freeze. The shock forced the breath out of my body. My heart seemed to stop beating. I felt crushed, unable to think. I had *rigor mortis*, without the *mortis*. And then I surfaced, spat out the dirty water and, gasping for breath, found my voice. 'Stop!' I shouted. 'Don't listen to me! It's bloody awful! Stay where you are!'"

The brutal cold ensured that when the rest of the cast followed More into the pool, their scramble for lifeboats did not require acting.

At eighty, Lawrence Beesley would have fared poorly in such cold water, just as how forty-six years earlier, in 1912, he likely would have perished with the others. Back then he was saved by luck, and this time by the grace of a technicality. After he gate-crashed the set earlier that evening, volunteering to go down with the ship, his presence prompted a meeting of the film's director and senior executives, who agreed that, as nice as it'd be to have an actual *Titanic* survivor in the film, the production was strictly a union affair. Beesley lacked membership in the actors' union, and so he was permitted to watch the filming. After that, he was again spared the indignity and denied the glory of going down with the ship, and was shown the door.

When it appeared in theaters in 1958, *A Night to Remember* became the seminal record and most definitive account of the real event. Critics gave it unanimously rave reviews for its inventiveness and storytelling. Its script wove together dozens of vignettes, giving the viewer the sense they weren't just following one or two characters, but were instead in every part of the ship as it filled with water. The lack of big-name actors and fictional romances formed an accurate retelling of what was ultimately a heartbreaking tragedy. Perhaps as a consequence, the film never saw huge crowds in theaters. In black and white, it did significantly better in England than in America, which had already moved mostly to color and actor-centric films driven by names like Jimmy Stewart and Marilyn Monroe. *A Night to Remember* made the ship the main character and eschewed any extra embellishment for a story that Walter Lord, who had written the book that inspired the movie, said was so powerful it "needed no added drama."

But even with mixed commercial success, Lord's book and the film that followed had an impact nobody expected. After more than forty years, Walter Lord had almost single-handedly revived the story that was falling further and further into the fog of long-ago events, and he renewed the elements of its inherent melodrama with astonishment, fascination, and obsession.

Doug Woolley saw *A Night to Remember* like thousands of other Brits, in a small theater in central London after work one day. While he watched the story of desperation and death, he couldn't help but grin at seeing his favorite story on the big screen. He went back to see it again and sat in the front row, as close as he could get to feeling like he was on the ship. If George Woolley had enchanted his grandson's imagination as a boy, then by the late 1950s, Walter Lord unleashed it with compulsive intensity as a man.

---

Surrounded by news clippings and old photos, Woolley was flummoxed by the magic of the story. The same elements of the disaster that had drawn mass public interest in the days of April 1912—the largest ship, the richest passengers, the worst corporate failure—had grown more powerful through the lens of time.

This wasn't a slow growth of the *Titanic*'s legacy so much as a revival. After the tsunami of *Titanic* news crashed in 1912, the story of the ship ebbed into a slow-moving stream. Headlines about hysterical last moments and lost husbands devolved into legal squabbles about insurance payments and shipping regulations. Like any former A-1 story, the growing distance from the event pushed headlines about the tragedy to pages eleven, then twenty-two, and then off the papers entirely.

As had been demonstrated by Charles Smith, obsessing about an aging shipwreck was frivolous in the 1910s, particularly when one looked around the world and saw much higher stakes and more lives at risk in war than had ever been threatened by an iceberg. The *Titanic* had also been eclipsed by an endless parade of new wrecks with high casualties. A year after the *Titanic* sank, an inland hurricane in the Great Lakes wrecked a dozen freighters and killed more than two hundred people. A year after that, the passenger liner RMS *Empress of Ireland* was accidentally rammed by a cargo ship on the Saint Lawrence River in Canada. Even though it had enough lifeboats, the *Empress* sank in just fourteen minutes and killed more than a thousand people.

Despite well-publicized improvements made to shipbuilding and emergency preparedness after the *Titanic*, maritime disasters in 1915 started to speed up. On July 24, 1915, a passenger liner in Chicago named *Eastland* sank because she was carrying *too many* lifeboats and lifejackets, which made the top-heavy ship roll over. Eight hundred forty-five people were killed. That summer, the early days of World War I began to yield an average of six wrecks per day, most of them British and the vast majority at the hands of enemy torpedoes, shells, or scuttling. There also continued to be more civilian accidents than anyone would expect, some just days apart, like the American schooner *Crockett*, which was toppled by a hurricane in Galveston, Texas, on August 17; the British navy trawler HMT *Poonah*, which collided with another vessel near Turkey on August 18; and the Brazilian passenger ship *Orion*, which ran aground near São Paulo on August 21. Altogether August 1915 posted one of the worst months for major shipping accidents ever recorded. In all, one hundred seventy were sunk, grounded, or otherwise lost without record.

The quick decline of the *Titanic* in popular culture of the 1910s had proven that the disaster had become a commodity, a means to sell books, newspapers, and at least one early film, *Saved from the Titanic*, released exactly four weeks after the sinking, before public interest waned. The Travelers Insurance Company used the *Titanic* to rake in money selling insurance policies for travel on the high seas, and theater owners sold tickets to public showings of photographs assembled from the deck of the ship. Many of the images were fraudulent or obvious forgeries, but audiences came all the same.

Once the commercial appeal faded, the *Titanic* limped along through the decade and into the 1920s through a series of memorials and remembrances. First Lady Helen Taft, whose husband had been president when the *Titanic* sank, raised money for a memorial on the banks of the Potomac. Every year for more than a decade after the disaster, the Coast Guard sent a ship to the sinking coordinates and played taps. For a while, the *Titanic* became a focal point for all maritime tragedies. Each April, sailors and seamen of any kind assembled at the Seamen's Church Institute in lower Manhattan for a vigil for everyone ever lost at sea.

In the years after the sinking, the *Titanic* began to mean different things to different people, and it took on an unexpected racial component. Newspapers that served Black Americans in New York and New Jersey largely glossed over the disaster. Just one passenger on the entire *Titanic* had been of African descent, a Haitian man named Joseph Laroche, who had put his wife and daughters in a lifeboat and then went down with the ship. But as time went on, Black Americans sought to claim a role in the cultural retelling of a story they had been expressly excluded from. Religious African Americans saw the *Titanic* as an example of God's intervention in concentrated wealth and knowledge among whites. But secular Black Americans responded to the sinking with the same sort of gallows humor and whispered sarcasm that had served their parents and grandparents in slavery. Several years after the sinking, African Americans began telling the story of Shine, a fictional black stoker aboard the ship who survived because of his foresight, courage, and stamina, which, it was implied, the white passengers lacked. In the story, which became a poem delivered as a toast, Shine jumps off the ship and swims to shore, outswimming a whale along the way, and takes up in a bar and starts drinking. While the world is learning about the disaster, Shine is regaling other drinkers with tales of his heroics. With every retelling, the words changed slightly until the 1950s, when Langston Hughes invited readers around the country to send him their own variations of the Shine story, a collection of which were published in the 1974 book on Black folklore tradition, *Get Your Ass in the Water and Swim Like Me.*

For Asian passengers, the experience was the opposite. Rather than write themselves into the story they had no part in, six Chinese survivors had to fight for decades for the basic acknowledgment that they were in fact aboard the *Titanic* and survived. One of them actually beat death by clinging to a door in the water, exactly like the character Jack in the 1997 film; he was picked up by a lifeboat. As told in a 2020 documentary called *The Six*, about the Chinese men's almost endless saga, the six men were contracted to work on other ships leaving from America, but when their transport to the United States sank, they instantly became subject to a racist government policy. The Chinese Exclusion Act forbade Chinese immigration to the U.S. (a policy in place until 1965), so when they arrived in New York on the *Carpathia* and the rest of the nearly seven hundred weary survivors disembarked and were swarmed by reporters and aid workers, the Chinese men were prohibited from touching U.S. soil. They spent that night aboard the *Carpathia* in harbor, and the next day, they were deported to Cuba.

Plenty of others found the *Titanic*'s reputation objectionable, even insulting. Its themes of class privilege and the oversize bubble of news coverage underscored the widening class divide. "Had the *Titanic* been a mudscow with the same number of useful workingmen on board and it had gone down while engaged in some useful social work the whole country would not have gasped with horror, nor would all the capitalist papers have given pages for weeks to reciting the terrible details," wrote the Kansas working-class newspaper *Appeal to Reason*. Another paper, the *Railroad Trainman's Journal*, pointed out that half a million people died each year in the "steady grind" of collapsing mines, train wrecks, sweatshops, and from tuberculosis. The *Titanic* wasn't only a small tragedy in comparison—the fact that anyone cared at all was an insult to working-class people who toiled every day in dangerous conditions and lifelong obscurity.

And yet, the ship that died a waterlogged death refused to be killed above the surface. Every so often through the twenties and then the thirties, the *Titanic* would appear again, slowly transforming into metaphor. Every historic event seemed to point the lens backward to the *Titanic*. The First World War was a sign that people weren't supposed to mix cultures or travel across oceans, such as on the *Titanic*. Charles Lindbergh's 1927 flight across the Atlantic had demonstrated that aviation was the future and that *Titanic*-like shipping disasters were relics of the past. The Great Depression showed that corporate power had grown too large and unchecked, and the *Titanic* was proof of its disastrous effects. Seemingly any argument could include the phrase "But what about the *Titanic*!" And the arguments, like the retelling of details, would circle around and around.

And no one cared more than Doug Woolley, who almost from birth intertwined himself with the *Titanic*. In any spare time, Woolley imagined himself on the ship. He imagined conversations he would have during its final moments. He would see every film ever made about the disaster and hoard every letter, poster, and newspaper clipping that mentioned it.

After long enough, it was only reasonable that the papers started to mention him, too.

° ° °

Trust was the rarest commodity in Doug Woolley's life. His mother had mocked him, his friends abandoned him, and his co-workers always wondered what he was up to. He felt like he was always being followed and that anytime he wrote something down, he'd have to tear it up and burn it. The only person he could trust was Pops, his grandfather, who was now an old man.

Doug's eccentric personality made it difficult to hold down a job. In 1951, he got a position answering phones at the Stroud Green Telephone company in Haringey. He stayed about six months before he was shown the door, and then he found a job in a sheet metal factory in Knowsley, where he carried pieces of steel from one machine to the other. This lasted for nine weeks before he was urged to apply at the Dunlop Rubber Factory, where he would help put together tires. Two weeks later, he was out of work again and fell in with a local farmer, who turned Doug toward fields of beans and potatoes and told him to talk to himself all he wanted while he was out picking.

Most of what Doug wanted to talk about was how to pull a ship out of the water. When he was a teenager in the late 1940s and early '50s, the *Titanic* had been something of Woolley's secret, a historical event that everyone knew but one he knew intimately, like being close friends with a celebrity. He could rattle off trivia about passenger lists and dining menus, and, in his time alone, he carried on elaborate interviews with the ship's captain, Edward Smith, about the liner's speed and nighttime navigation. *A Night to Remember* gave Woolley rich new details about the characters on board and the final moments of mayhem. But it also lit a fire in him—if the *Titanic* was making a pop culture resurgence, he'd need to move faster.

By 1960, something about the *Titanic* story took off. A Massachusetts jeweler named Edward Kamuda founded the Titanic Enthusiasts of America, the first known fan club for a shipwreck. Word spread, and other chapters popped up, in Chicago, New York, and Phoenix. The groups were almost exclusively men who sniffed brandy and discussed details of the famous story and the what-ifs and what-might-have-beens. Kamuda wrote to all eighty-seven living survivors and invited them to join the organization. Seventy-five agreed. Within a few years, the Titanic Enthusiasts of America had gone international, and Kamuda—having received more than one complaint about the awkwardness of being enthusiastic about a tragedy—changed the name to the Titanic Historical Society.

On the other side of the Atlantic, Doug Woolley was less interested in survivors and fan club chitchat than in the waterlogged ship itself. What kind of condition was it in, and how could he get close enough to take a look? Without any evidence and based on largely debunked rumors, he believed there could be more than ten million pounds in hidden treasure and valuables, including the combined wealth of ten millionaires on board, a De Beers diamond, the black Buddha crown stolen from the Dalai Lama, seven parcels of holy parchment from ancient Israel, and a bejeweled papal cross worn by the pope himself.

But what gnawed at him most when it came to the wreck of the *Titanic*, was who *owned* it?

The question was steeped in thorny legalities. Any ship is owned by the person or company that paid to build it. In the *Titanic*'s case, this was the White Star Line, which, upon payment of its $7.5 million price, owned every rivet, boiler, and four-post bed on the ship. But when it sank, its contents were not only substantially devalued, they became the property of a series of Lloyd's of London underwriters that had insured less than half of the ship's value, a small fraction that still seemed ludicrously high for a ship considered unsinkable. After the underwriters paid out insurance claims and settlements, Lloyd's technically owned what was left of the ship. But with little confirmed value on board, executives at Lloyd's were eager to disassociate from the tragedy and write off the loss. Beyond its cultural value and any salvageable artifacts, the wreck itself was financially worthless.

This gave Woolley an opening. By 1962, he was working as a signalman in Welshpool pulling levers to direct trains to one track or another. During hot afternoons sitting in the signal box, Woolley daydreamed about how he could find the owners of the *Titanic* and convince them to let him salvage it.

One day after work, he wrote to a maritime lawyer in London. A few weeks later, the lawyer wrote back, informing Woolley that if he really wanted, there might be a way the *Titanic* could be his.

° ° °

Owning the most famous shipwreck in history would be exciting enough. But Woolley had a bigger plan once the ship's deed was awarded to him.

Establishing ownership was relatively simple, the lawyer told him. First he would have to verify that no one else owned the wreck of the *Titanic*, which was a hazy legal question. The insurers had disavowed it, and no one else had made an obvious claim for it, which was enough to convince Woolley there was no owner. From there, to say Woolley did his due diligence sorting through every document, witness, and relevant court case would be like saying a schoolboy thoroughly investigated how to get himself elected president. He read several books, made perfunctory calls to Lloyd's of London's front office, sent several letters to the British Board of Trade, and then mostly called it a day.

Maritime law has an intricate and incongruous way of handling ownership of sunken ships. Vessels built by governments, such as battleships or destroyers, are the property of those countries, whose courts almost never give up salvage rights. But private vessels are subject to flimsier rules. Because ships are inherently mobile and spend much of their time in international waters, the question of their jurisdiction can change quickly. If a ship built in America but registered in Jamaica embarks on a voyage from Brazil to Ireland with mostly Italian passengers on board and sinks in international waters, which country's courts have the power to make judgments over liabilities and any salvageable cargo?

"Ninety-nine percent of the time, nobody wants the bloody thing," Nick Gaskell, an Australian maritime lawyer, told me one morning. "It's a hunk of metal at the bottom of the sea and does nobody any good."

But that leaves the 1 percent of ships that aren't worthless. Perhaps they were carrying precious cargo or had expensive equipment, such as the famous case of the SS *Central America*, a steamer carrying almost six hundred passengers as it traveled from California to New York City in 1857. Americans mourned when it sank in a hurricane off the coast of the Carolinas, but the more lasting effect was economic pain. In addition to the gold miners and wannabe prospectors on board, the ship was carrying three tons of gold worth $8 million, the modern equivalent of more than half a billion dollars. American banks in New York were awaiting the gold to provide liquidity to the economy, and when it disappeared six thousand feet underwater, the nation entered a financial panic that lasted almost a decade.

After paying insurance claims for the wreck, sixteen insurance companies in the U.S. and the UK owned the *Central America* and its contents. They never expected to get it back. But in 1988, when a salvage firm announced that it had found the one-hundred-forty-year-old ship, half of the insurance companies still existed and, using newspaper clippings and corporate minutes, they proved the gold belonged to them. The head treasure hunter, Thomas Thompson, sang the blues in front of a judge: "It saddens me to think that it could be possible that a group of the largest insurance companies in the world could be awarded the treasure we worked so hard to find and recover." Thompson ended up with $50 million to distribute among the dozen investors who underwrote the salvage operation. But instead, he kept the money for himself, left town, and was pursued by the FBI until he was caught in 2015. A collectibles expert told me that gold pieces from the *Central America* still show up at auctions every now and again. "For anyone who wants to own one, it's not hard," he said.

Salvage law is written to encourage operations like Thompson's. If someone is willing to go through the effort to find a wreck, the law generally agrees they deserve a portion of its value. The same is true for ships in distress. If a vessel is at risk of sinking or blocking traffic, a Good Samaritan can help, and if the salvor saves the day, a court will often grant them a generous reward. Such rewards can be expensive for the distressed captain, but several maritime lawyers told me that the payments incentivize people to help others in dangerous environments without obvious emergency response.

The law of *finds*, however, involves a sense of discovery—a finders keepers of the sea. If a wreck has been abandoned for a number of years and no owner or insurer has made any effort to find or reclaim it, then the wreck enters a legal purgatory. This leaves it open to scavengers and shipwreck enthusiasts. Convince a judge in the logical jurisdiction that you are willing to put in the effort to do *something* with the ship, usually by proving that you actually have been to the ship's location and have brought back a piece of it as proof—an artifact known as a res, from the Latin word for "thing"—and if no one objects, the ship can be yours.

This was the provision of law that Woolley sought to exploit. By 1965, Woolley reasoned that the *Titanic* had been abandoned for more than fifty years. No one even knew where it was. Since Charles Smith's failed efforts in 1914, not a single person had visited the wreck or had any serious plans to move it. The White Star Line had gone belly-up in 1934 and then merged its defunct assets with Cunard, which then erased virtually all mention of White Star by 1950. Cunard wasn't going to fight for an aging wreck that it never even owned, and Lloyd's of London had long ago washed its hands.

Woolley thought that if he could cobble together a few thousand pounds, a round and not paltry amount, Cunard might write him a letter saying, sure, it's yours. But even this plan failed when Cunard rejected Woolley's correspondence altogether, saying they wanted no part in the *Titanic*. A junior officer at Lloyd's echoed the sentiment that as long as the ship sat underwater, it was worthless. The same thing happened when Woolley sought waivers from several descendants of victims who had property on board.

Woolley's low-cost lawyer gave him more budget-friendly advice. He told him not to bother with going to the wreck and the formality of plucking off a res. This was convenient, since Woolley, who was working in a factory making nylon pantyhose, had neither the means nor the know-how to even get to where the *Titanic* sank, let alone miles underwater. All he had to do, the lawyer said, was place a classified ad in four or five newspapers announcing that he was claiming ownership of the wreck of the *Titanic*. He advised Woolley to do it on a holiday, when fewer people would be reading, and the smaller the font the better.

Years later, Woolley would say he did exactly this. He wrote to the *Coventry Evening Telegraph* and the *Newcastle Evening Chronicle* and submitted three-line classified ads stating that he, Douglas John Faulkner-Woolley, age thirty-one, laid claim to the salvage rights of the *Titanic*, former ship of the White Star Line. According to Woolley, he included his name and address and invited any objectors to respond within two weeks. Then he recalled that he sat nervously by his postbox every day waiting for legal papers or stated objections. Hearing none, Woolley was told by his lawyer that he could consider the *Titanic* functionally his.

In reality, Woolley appears to have skipped every step but the last one. I was skeptical that this was how shipwreck rights changed hands, even in the days before the internet. I also had trouble believing that such a brash classified ad about the world's most famous wreck would be met with complete silence for two weeks. Perhaps people thought it was a joke and turned the page. But this was 1967, a time when everyone read newspapers, including classifieds, and when anything that mentioned the *Titanic* had become fresh and exciting. I consulted several maritime lawyers and historians, who believed it far more likely that Woolley never placed such classified ads and merely claimed that he had. And, as it happens, a search through large and small British newspapers yielded no matching classified ads in the 1960s—the era when Woolley was maneuvering to launch his master plan—no mention of ownership rights, and no invitation for anyone to object. Instead, it appears he simply began telling co-workers, friends, and anyone he met that he owned the most famous shipwreck on earth, a claim tailor-made to dominate dinner parties and workplace chatter. "I regard the *Titanic* as mine," he began saying.

This was a tall tale, but not an obvious lie. Lying requires knowing the truth and rejecting it, but the more he repeated it, the more Woolley seemed to really believe it, and the more confident he grew, the more convincing it became to everyone else. "I'm the only person to have filed formal legal claims, so I reckon the *Titanic* is mine," he said years later, summing up his argument. If the *Titanic* effectively belonged to no one, then why shouldn't it belong to him? And if it could, then he concluded that it did.

○ ○ ○

While still out of reach, the ability to visit the *Titanic* had taken leaps by Woolley's era. In 1934, the famed scientist William Beebe unveiled a steel ball that would allow two people to investigate the sea as deep as three thousand feet, which amounted to a half mile down and just shy of a quarter of the distance to the *Titanic*. Beebe designed the device, called the bathysphere, to mimic the underwater homes of a spider known as the "diving bell spider," which weaved and spent their lives in waterproof cocoons. The spiders were a marvel, and Beebe reasoned that if spiders could live underwater, so could humans. He spent three years strengthening the steel sphere from leaks and cable breaks likely to arise under seven thousand tons of pressure. When he began experiments, the overwhelming sensation of being so deep was the darkness, a color "blacker than black," he recalled. And even amid constant mishaps, including once when the crew on the boat above watched a rope snap and believed the bathysphere and its occupants were lost to the ocean forever, Beebe continued to go back.

A decade later, underwater explorers Jacques Cousteau and Émile Gagnan perfected an underwater breathing device called the aqualung, which allowed a person unconstrained by steel or cable to descend to a depth of about one hundred feet. Any diver who went that deep quickly felt the effects on his body and reported to those on the surface why sending a human down thousands more feet was preposterous. Their breathing device wasn't the first, but it was the most popular, which expanded ocean exploration from the niche domain of scientists to the general public. Early encounters of underwater life spread into pop culture, including the peculiar 1956 blockbuster *Manfish* and the weekly TV show *Sea Hunt*, about a detective who solved crimes and fought bad guys below the waves.

The most significant advance in this golden age of underwater exploration came from the brain of Auguste Piccard, who had ironically built his expertise as a hot-air balloonist. Piccard had taken large helium balloons to record-breaking altitudes in the atmosphere in the 1930s. He used the same design to construct what he called a bathyscaphe suspended by large floats overhead. Like a submarine, the floats regulated the ratio of water to air, which allowed the craft to control its own buoyancy and navigation without the assistance of a tether. In 1960, Piccard's son Jacques took the most elaborate bathyscaphe ever built to the deepest point in the ocean, a seven-mile-deep depression in the South Pacific known as the Challenger Deep.

Piccard's bathyscaphe for that mission, the *Trieste*, a craft that looked like a giant slug with zebra stripes, was so large and heavy that it couldn't sit aboard a ship but instead had to be towed in the water. It also required an enormous amount of energy to make the fourteen-mile round-trip descent, burning through its fuel tank of forty thousand gallons of gasoline. In a time before satellite GPS could pinpoint the precise site of the deepest crevice, the team found the dive site the old-fashioned way, by tossing TNT over the side of the ship and timing the low-hummed echo that reverberated down to the bottom and back up to the ship's hydrophone.

On January 23, 1960, going down was a triumph, but unlike Beebe in 1934, who marveled at every fish, eel, and squid he could see swimming by the windows, the descent on the *Trieste* was more symbolic than visual. The craft had a single window the size of a quarter (anything bigger risked the structural integrity of the craft) and the view was either pitch-black or hazy with mud. "It was like looking into a bowl of milk," Don Walsh, a navy engineer who made the descent with Jacques Piccard, later said.

Meanwhile, in the Atlantic, geologists were using the latest renderings of the deep sea to decipher earth's history. The American geologist Marie Tharp studied sonar profiles in the mid-Atlantic and noticed a curious seam, almost like the suture of a peach, that extended thousands of miles from Iceland to Antarctica. Her studies revealed a rift stretching more than forty thousand nautical miles, a discovery that helped formulate theories that the continents were not stagnant but in constant motion, crawling millimeters every year. Extrapolating the implications helped explain the continental stress that causes earthquakes.

This type of emerging knowledge of the deep sea formed the foundation of a new industry devoted to salvaging ships. Many young children love ships, but it was the rare one in 1924 who became obsessed with ship demolition and wreck removal, a niche but lucrative line of work pioneered in England by a young man named Risdon Beazley. Beazley started demolishing ships as a teenager and by age twenty-two started a company to professionally destroy wrecks. Owing to his unmatched experience, Beazley won lucrative contracts from the British government to get rid of junk ships, starting with small boats in shallow water and growing to larger projects that required so much advanced equipment that he had almost no competition. In 1937 alone, he completed two enormous excavations. He cut up and disposed of the cargo ship *English Trader*, which had struck ground in Dartmouth Harbor, and pulled up and salvaged the *Kantoeng*, then the largest dredge on earth.

Ships were salvageable if they had sunk relatively recently. But Beazley began dreaming of older ships, ones long faded in the world's consciousness. By the 1950s, he shifted his focus from ship salvage to cargo recovery, which would be considerably easier to execute and net him far more money. The most obvious ship to target was the *Titanic*; not only did it carry valuable items, but on account of its celebrity, every fork, doorknob, and thimble he could surface from the wreck would sell for many times its original value.

Beazley launched the first serious attempt to salvage the contents of the *Titanic* in July 1953. Unlike Smith or Woolley, who saw their best chance of success by whipping up a frenzy, Beazley told no one when he chartered the Admiralty Coastal Salvage vessel the *Help* to carry out a secret mission, which was intentional, since Beazley had no intention of tiptoeing around a fragile cultural artifact that also doubled as a grave site. Instead he dropped explosives above the presumed site in hopes of hitting the wreck and blowing open the hull so he could retrieve objects inside. Almost anyone alive would have considered this shockingly reckless. But Beazley was playing for an audience of one: himself. Luckily, he had the wrong site by several miles and ended up dropping TNT onto seafloor mud. He tried again a year later but got no closer.

Eventually Beazley would realize his folly and ineptitude. He would mature from a naïve treasure hunter into a serious wreck explorer who leveraged experience and built partnerships. In a masterstroke, he built two salvage vessels, the *Twyford* and the *Droxford*, and got the British government to help pay for them under a vague agreement that the Crown would be entitled to a portion of whatever cargo he recovered. This worked out for all parties when, by 1965, he'd pulled up more than fifty-five thousand tons of precious metals, including copper, nickel, lead, and, of course, gold.

With the secret out that wrecks were one of earth's great untapped sources of wealth, the elbowing sparked a literal gold rush of the sea. The process was haphazard but designed to be quick. Treasure hunters compared maps of the seafloor with famous trading routes of past empires. Find a spot of shallow water and pack your things to leave the next day. No one knew how much precious metal might sit underwater, but whoever got there first would find out.

No place on earth was more popular than Florida and a stretch of shore known as the Treasure Coast, where a dozen Spanish ships were returning from Cuba to Spain in 1715 with thousands of ounces of silver when they were struck by a hurricane and sank. In the summer of 1967, so many hunting boats bobbed off Florida that they sometimes bumped into one another and captains employed armed guards on their decks to keep the peace. On an average day that summer, two hundred sixty-four pounds of silver were reportedly pulled out of the water. Some people celebrated publicly, but not all. Amid fear of piracy among neighboring boats, others denied they found anything at all.

These were the days before the U.S. began to regulate lost treasure and salvage rights, an era of dumpster diving at sea that yielded incredible reward. Mel Fisher, a Florida man who lusted for and eventually found the most heavily laden of the Spanish ships, the *Nuestra Señora de Atocha*, a ship believed to be carrying $400 million in silver, gold, copper, gemstones, tobacco, and indigo, had two mottos he repeated every day he was on the water:

*Today's the day* and *Finders keepers.*

Chapter 7

# BATHTUB EXPERIMENTS

Declaring his association with the century's most famous disaster brought Doug Woolley a sense of credibility and even mild celebrity. No longer was he just employee number 64 at the Hertfordshire hosiery factory. Woolley was a maritime guru with an expertise in ship salvage who was recognized at the pub and invited to speak at universities about his *Titanic* ownership and his plans to bring it back. Every ounce of attention seemed to beget more, and Woolley, who had seldom even *ridden* on a boat, fielded regular visits from reporters who sought his expertise on matters of deep-sea engineering.

Woolley enjoyed these visits and relished the opportunity to be quoted so much that he often invited the newspapermen inside and put on a pot of tea while he offered a tour of his endless memorabilia, which supported his persona as a credible maritime expert. It was a feedback loop, and it worked perfectly.

Woolley's original plan was almost identical to Charles Smith's, with some refinements in measurement. It called for two hundred pontoons evenly placed around the hull of the ship, all connected by nylon rope. He calculated that the *Titanic*'s deadweight was roughly sixty-six thousand metric tons, which meant each pontoon would be responsible for just over three hundred tons—heavy but doable. Once the pontoons were in place, ships on the surface would start to generate twenty megawatts of electricity to extract eighty-five thousand cubic yards of hydrogen from the seawater via electrolysis and pump the gas two miles underwater to inflate the pontoons. This assumed that the pontoons could anchor themselves under the ship, so as to propel the ship from below rather than pull from above and rip it apart. Woolley thought of a solution here, too. The same electrical system that would shoot the hydrogen underwater would create a powerful ultrasonic boom to separate the ship from the mud. It would be quick, maybe two or three seconds, and he'd have only one shot. The shock wave was designed to hit the ground only, but Woolley hoped that the boom didn't accidentally damage the ship itself, or what was left of it.

In August 1968, Woolley revealed this plan in a press conference he held in his apartment. He said it would cost £3 million, which was more money than Woolley had ever imagined existing in one place, let alone under his direction. But he also announced he had a backer, an electrical engineer he knew who Woolley said had promised "some money" to set up an experimental salvage company.

After the articles ran, Woolley began receiving letters from actual engineers and wannabe collaborators. Woolley sorted through them with amusement and almost zero discretion. In Woolley's view, anyone who wanted to help was qualified for the job; the bigger the team, the more senior his role as kingpin. By the spring of 1969 he

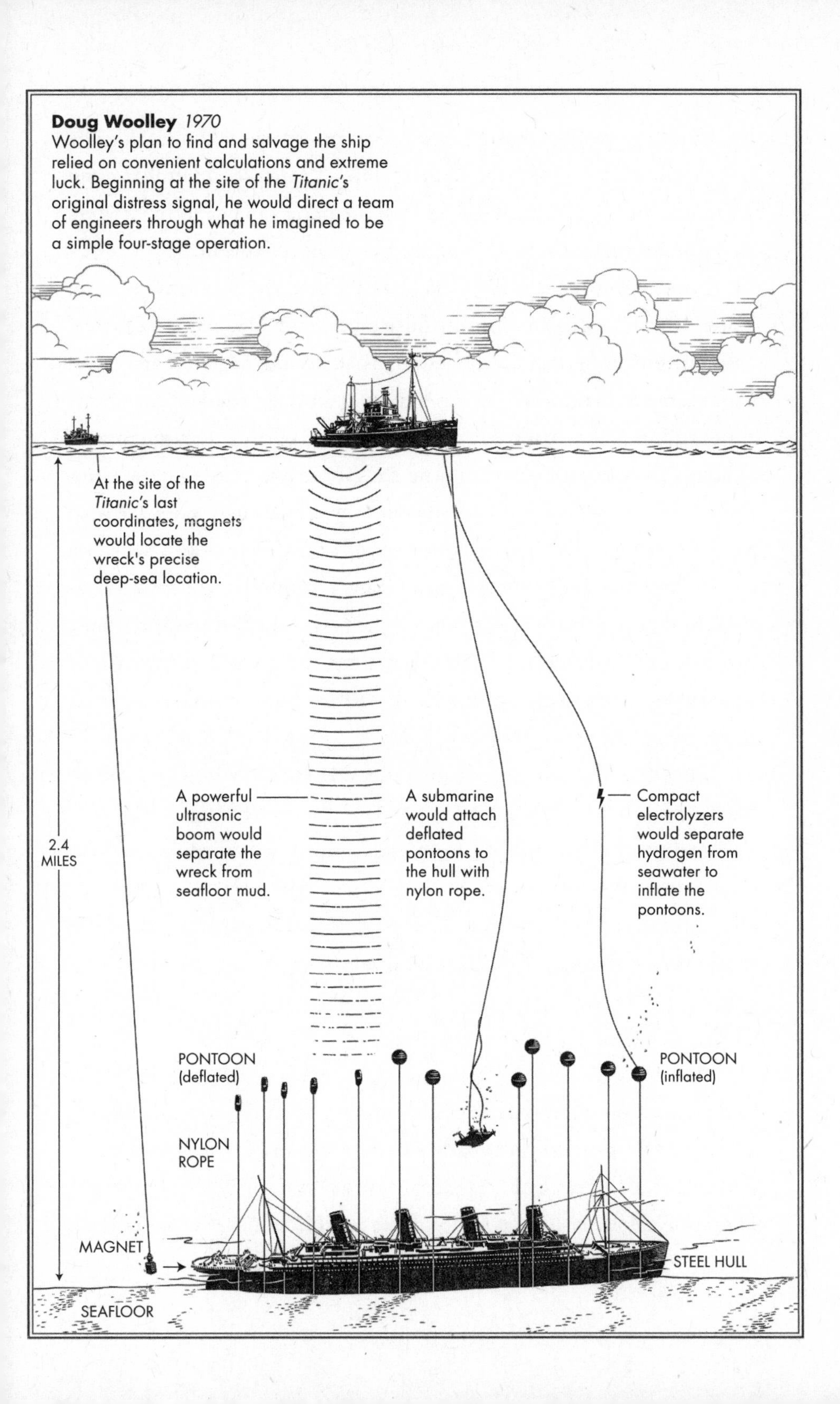
**Doug Woolley** *1970*
Woolley's plan to find and salvage the ship relied on convenient calculations and extreme luck. Beginning at the site of the *Titanic*'s original distress signal, he would direct a team of engineers through what he imagined to be a simple four-stage operation.
At the site of the *Titanic*'s last coordinates, magnets would locate the wreck's precise deep-sea location.
2.4 MILES
A powerful ultrasonic boom would separate the wreck from seafloor mud.
A submarine would attach deflated pontoons to the hull with nylon rope.
Compact electrolyzers would separate hydrogen from seawater to inflate the pontoons.
PONTOON (deflated)
PONTOON (inflated)
NYLON ROPE
MAGNET
STEEL HULL
SEAFLOOR

began telling people that he had “gone international,” meaning he had received letters from the United States, France, and Hungary and was building a team, casually adding the names of the letter writers to publicity materials to inflate the size of his coalition.

Under scrutiny, it may have appeared that Woolley was running a con. Not only was he claiming false ownership of a major historical artifact, but he was selling the idea that he had the authority and capability to retrieve it. The operation resembled the famous scams of legendary New York con artist George Parker, the man responsible for the old joke about selling the Brooklyn Bridge. Parker not only sold the bridge multiple times to gullible customers, he also forged official-looking documents to sell the Metropolitan Museum of Art, the Statue of Liberty, and Grant’s Tomb. People love to believe more than they hate to be swindled, and the buyers of Parker’s bridge often refused to admit defeat until they were, in the case of one buyer, arrested for trying to install tollbooths on the popular roadway leading into New York City.

Woolley, however, was dead serious. If he was running a con, it was simply a Ponzi scheme of enthusiasm, reinvesting deposits of fascination from one person to the next, until he had spun a web so complete that he found himself caught in it. Not only was he personally committed to raising the *Titanic*, a crowd of people around the world was waiting for him to get it done.

° ° °

Every deck on a ship has a name. Starting from the top, the uppermost deck is called the weather deck, followed by the main deck, and then the upper deck, the deck best known for where Leonardo DiCaprio declared he was king of the world. Under the upper deck comes the middle deck, and if there is no middle deck, then one would descend straight to the lower deck, home to berths, staterooms, and cabins. Somewhere in this collection of decks might be a promenade deck, where one could stroll around the superstructure of a ship.

Depending on its purpose, a vessel can have any number of additional decks. The gun deck, helicopter deck, and hangar deck need no explanation. Cargo ships have tween decks between other decks to store weight low and centered. The most kid-friendly deck is the lido deck, usually home to swimming pools and water slides.

And finally the poop deck. For all nautical terminology's high-class veneer, at some point every captain, commodore, and midshipman eventually must reference the lowest brow but highest deck on the back of a ship, whose name comes from a Middle English translation of a Middle French interpretation of the Latin word *puppis*, meaning the stern of a ship. The *Titanic* had a poop deck, and it was where Kate Winslet almost jumped in her almost-suicide scene. But passenger ships aren't often built with poop decks anymore. The real estate is too valuable, and cruise vessels long ago converted poop decks to first-class cabins with sprawling balconies and seafood buffets.

The *Titanic* was built with ten decks divided into cabins, social areas, and cargo. The leisure space was interrupted by four boilers, whose smokestacks extended through the interior of the ship, up through the weather deck, and into the air. The mix of compartments and columns is what rationalized the term *unsinkable*, under the assumption that if water flooded in, the affected compartments would seal and the ship could sail on. Having neutralized water as a threat, few could imagine that the ship's fatal blow would come from the ship itself. The shear stress that ripped the hull in half was enough to rip through all ten decks and tear apart the first-class lounge, the reading and writing room, the promenade deck, the bridge deck, the shelter areas, the saloon, a group of third-class cabins, and more than three decks stuffed with enough coal to get to New York.

As the ship settled on the seabed, its decks were still largely intact. Columns had snapped and steel had buckled, but the dining rooms were still dining rooms and second-class cabins were still second-class cabins. The ship had lost its functionality as a ship and some areas were irretrievably destroyed, but other parts largely retained their structure. A deep-sea creature capable of surviving such conditions could still wander from room to room, checking out the wood carvings and the imploded furniture.

Each wreck is a time capsule in the process of falling apart, but until it does, you can swim through the hallways and lie in the bathtubs, which is the main draw of wreck diving. After it was struck by a mine in 1942, the SS *President Coolidge*, an art deco passenger liner turned World War II troopship, took its ornate decorations and twelve thousand tons of government cargo to the seabed off the island of Vanuatu. Eighty years later, every day of the year, divers swim through the cargo holds, the engine room, the galley, and the infirmary. They swim ironic laps in the underwater swimming pool and gawk at the classic chandeliers and mosaic tile. Far and away, the most remarkable part of the *Coolidge* is its accessibility. Stuck in relatively shallow water with easy beach access and high underwater visibility, the *Coolidge* is a guaranteed marvel. Thousands more ships offer equally jaw-dropping views of former eras if divers could reach them.

Wrecks disappear often one deck at a time. By 1968, it's likely that the *Titanic*'s ten decks had become six or seven as the structure slowly pancaked its weakening steel. No one was there to witness the moment the weather deck or poop deck collapsed onto the decks below, but the corrosion scientist Ian MacLeod thought that the first fifty years after the sinking were kinder than the next fifty years. MacLeod has spent four decades studying how ships break down by measuring the chemicals swirling around aging wrecks, and particularly the invasion of rust.

"One way to think about a wreck's devolution is to picture your house falling apart," MacLeod said one morning while I ate breakfast in California and he, in Perth, liberally sipped a late-night whiskey. He asked me to imagine if every year my walls got one millimeter thinner. For a long time I wouldn't notice, but eventually, the walls would be too weak to support the roof and the ceiling would fall in. A ten-story house would last longer, collapsing usually from the bottom up as the weight bore down, but with enough time and enough exposure to the elements, even the sturdiest house would be a flattened pile of rubble. And eventually, the rubble would be blown away, eaten by microbes, or swallowed by the earth, too.

After a half century, the two halves of the *Titanic* remained in decent shape. "A lot of the critical metal thickness that enabled the structure to stand was still there, so if it was possible to take photos in the sixties, they would have been considerably more evocative than the ones the world saw later," MacLeod said. "In the sixties, the two halves still had the appearance of a ship, and seeing them would have easily convinced the average person that it was still possible to raise the *Titanic*."

Raising it, however, would create a new set of problems. The same year Woolley was amassing influence and money for his beloved wreck, another British ship underwent almost the exact process Woolley had in mind.

When it set sail in 1848, the SS *Great Britain* was one of the first passenger steamers and the largest vessel ever built. It was deployed to shuttle wealthy passengers between Bristol and New York, and, with its novel steam-powered propeller technology, it was proudly seen as the arrival of a modern era. Unlike the *Titanic*, it survived its maiden voyage and made dozens more across the Atlantic. Sturdy and well-apportioned, the *Great Britain* spent thirty years carrying immigrants from England to Australia and then, after a fire, served as an immobile warehouse and coal storage ship in the Falkland Islands off the east coast of Argentina until she was unceremoniously scuttled and abandoned in 1937—an extraordinarily long working life of eighty-nine years.

The *Great Britain* would be forever submerged if a British millionaire who owned a football club hadn't happened to read about it in the newspaper one day. Wanting to preserve a symbol that had embodied England at its peak, in 1970, the millionaire pulled together a crew, rented a submersible pontoon to refloat the ship, and enlisted the help of a German tugboat to drag the hull eight thousand miles from Argentina back to England. It was showered with rose petals as it traveled up the river Avon to its original dry dock in Bristol.

But once the crowds left, engineers realized they had a major problem. While it was submerged, the *Great Britain* was rusting and falling apart. But once it resurfaced, it began to rust *much* faster from oxygen and humidity in the air and the salts embedded in every crevice. Left alone, the hull was in such poor shape that structural engineers gave it six months before it completely crumbled.

Rather than abandon an effort that had already cost millions, engineers constructed a giant dehumidification chamber for the bottom half of the ship that effectively preserved the hull in conditions comparable to the Arizona desert. The outer structure was filled with resin and coated in anti-corrosion paint. Inside, dining rooms were reapportioned with nineteenth-century furniture and passenger bunks were finished in new wood. Start to finish, the conservation took three years and £11 million. It was reasonable at the time to wonder if the effort was worth the price. The issue was settled, however, when the tourists began buying tickets to see it. Fifty years later, the SS *Great Britain* remains Bristol's most visited attraction. You can get married on it, sleep on it, or, according to its website, rent it to host your corporate event.

Like the *Titanic*, the draw of the *Great Britain* is the result of good storytelling. A formerly regal ship, lost and refound, returned to its home dock on the anniversary of the date it was launched. Were the *Titanic* built smaller or sunk in shallower water, one could expect almost this exact treatment to unfold in the early seventies, and today, tourists would flock to Belfast to see the *Titanic* the same way tourists gawk at Buckingham Palace or the Tower Bridge. Not a dollar or pound would be spared to return the ship's original sheen, and visitors would wander through its hallways with audio guides pressed to their ears recalling the stories of heroics and cowardice.

This shallow-water alternate reality would've been a done deal except for two wrinkles. One, unlike the SS *Great Britain*, the *Titanic* sank not by intentional scuttling but from a tragic accident that killed hundreds. Lifting the ship would require disturbing a grave with all the legal, physical, and spiritual red tape that entailed. And two, the *Titanic* would require a significantly more elaborate restoration, a race against rust on a larger ship that would deteriorate dramatically faster than engineers could preserve it.

This guaranteed notoriety for the person or people responsible for any screwup in salvaging the *Titanic*. Any weakening of the fragile structure, or, worse, a fatal miscalculation, would ironically leave one of the great relics of the twentieth century worse off above water than it was below.

"You'd go down in history as the person who accelerated the decay of the *Titanic*," MacLeod told me. "Who would want to be that person?"

° ° °

By the summer of 1969, Woolley was stretched thin. So many people were calling, writing, and offering to help him that he returned from work and stayed up sometimes all night answering every message. During the day, he was working as a clock winder at a medieval church in Baldock, which entailed the daily chore of climbing eighty steps up the church tower to wind it. Then he would walk to his other job as a dye operator in the pantyhose factory.

Something had to give, and after several months, Woolley decided to quit the church, and in quitting, he managed to squeeze a dollop of publicity for his *Titanic* plan. He called a local newspaper and told them the clock would soon stop because he had to devote his full attention to the *Titanic*. The paper took the bait and printed the story under the headline "Titanic Task to Stop Clock."

Woolley was also playing whack-a-mole with anyone who questioned him, or, worse, mocked him. While shopping for groceries one day, a man recognized Woolley from the papers and asked him how he could travel two miles underwater when no one had ever done it before. Easy, Woolley told him. The French already invented a bathyscaphe. His sister called one night and asked how he was going to get the money. Simple, he told her, the £5 to £7 million—a number based on Woolley's changing whims—would come from "people interested in the *Titanic.*" A Canadian newspaper sent a reporter all the way to England to meet Woolley and ask whether extensive damage might have made the ship unsalvageable. "In fact not," Woolley told him. "The lower depth the less oxygen you get and the less damage to the wreck."

Almost nothing Woolley said was based in fact. The French bathyscaphe was not suited to loading pontoons under such a sprawling wreck at such extreme depth. *Titanic* buffs would not line up to donate millions of pounds to a man lacking both credentials and experience in underwater engineering. And the bit about oxygen in the water column was at best only partially true. Oxygen is at a maximum at the surface and decreases significantly until about three thousand feet, but it begins to gently increase again with depth, a strange quirk of seawater that lets dissolved gases travel easily side to side but not up and down.

At home, Woolley concocted experiments to test his theory that he could inflate deep-sea pontoons. His idea centered on electrolysis, in which energy can split apart hydrogen and oxygen molecules of ocean water. If the hydrogen could be isolated, it could inflate the pontoons underwater and they would become buoyant.

Woolley worked on this experiment in his bathtub and spent several weeks building an oblong contraption capable of performing electrolysis. When it was ready, he affixed it to a balloon and turned it on. Hours later, the balloon had barely inflated. The next morning it appeared to have made no progress. At that rate, inflating hundreds of pontoons thousands of feet underwater would take millions of years.

Facts, however, didn't diminish Woolley's confidence that the details would work themselves out. Big projects always seemed impossible until they were inevitable. This was the same year Boeing unveiled the first jumbo jet and British Airways inaugurated the Concorde supersonic passenger turbojet, which could fly from New York to Paris at more than thirteen hundred miles per hour. Earlier that summer, two American astronauts walked on the moon for the first time, and that display of human capability enchanted the world. Woolley drew a direct comparison. "Of course I know it's an odd obsession," he said in October 1969. "But flying to the moon is an odd obsession as well."

Later that month, at the precise moment that Woolley had seemed to squeeze every drop of attention from a hypothetical idea lacking any forward motion, he announced that he had assembled an international team of experts. Two Hungarian engineers had written to Woolley, which, in his view, was enough to consider them on the team. There was an Austrian chemist who offered to help with the floating component. Two of his co-workers from the hosiery factory wanted in, along with a London accountant, two Massachusetts businessmen, and the proprietor of a local fish-and-chips shop.

With the exception of his hosiery friends and the fish-and-chips fellow, Woolley never met anyone on his team. All of the coordination was done through the mail. The Hungarian engineers wrote to Woolley to tell him they had "a machine" fitted with underwater lights and mechanical arms to manage the nylon bags. The Austrian fellow said he had done calculations about gas buoyancy or something or other. And the rest offered to help with publicity and fundraising the £4.8 million someone had suggested to Woolley as a budget. For something so uncertain, the whole affair in Woolley's mind was a done deal. All that remained was the most important detail of all: once the ship was surfaced and the brass, copper, and other valuables on board were sold, every member of the team would receive £600,000, a number big, round, and largely plucked from thin air.

Plenty of people declined to be on Woolley's team because they found the idea ridiculous—or, worse, they didn't care. But one notable absence was Woolley's own government. Despite well-publicized but misleading reports that "the Hungarians" were on board and "Austria" might pitch in as well, the British Crown laughed Woolley out of the room in the most British of ways: silence. He couldn't get a single letter answered by the Sea Transport Branch of the British Board of Trade or the Royal Navy or from Buckingham Palace itself.

This was ironic and, in Woolley's view, unfortunate because it was wholly reasonable that if a government as powerful as England's got involved, the *Titanic* could at least be visited, if not salvaged in small pieces. The price would be steep, and conservatively, the effort would require a navy. But Woolley was correct in believing that the technology, expertise, and appetite existed for nations to rescue old ships in the deepest parts of the ocean if they wanted to badly enough.

The proof was sitting, inconveniently, twelve thousand miles away from Baldock, England, in the North Pacific. A year earlier, a Soviet submarine carrying three nuclear-armed ballistic missiles sank about fifteen hundred miles north of Hawaii. The crew was lost, and the vessel sat under three miles of water, even deeper than the *Titanic.* The Soviets, not wanting their secret sub to be discovered, wrote off the loss as if nothing had happened. They might have gotten away with it if not for American intelligence analysts who eavesdropped on Russian shipping communications while the ship went down. If the Americans could recover the wreck and the undetonated missiles—and especially if they could do it in secret—the wreck would yield invaluable intelligence of Soviet capabilities at the height of the Cold War.

The Soviet submarine, the *K-129*, was smaller than the *Titanic*. But it rested nearly a third deeper, at sixteen thousand feet, in a remote part of the Pacific. Getting to it would be difficult, but resurfacing the entire seventeen-hundred-ton, 132-foot-long vessel would be far harder. In secret, the CIA spent two years planning a feasible approach: using a large mechanical claw suspended from a surface ship by enormous winches. The plan was not unlike Charles Smith's idea in 1914 or Doug Woolley's in 1969, except Project Azorian, as the top secret maneuver became known, was underwritten with almost $1 billion by the U.S. government. It was also cloaked under the plausible cover story of being an undersea mining venture ostensibly sponsored by the reclusive billionaire Howard Hughes. When visited by CIA agents, Hughes allowed the government to use his name on the deep-sea drill platform, the USNS *Hughes Glomar Explorer*. He consented to establishing an elaborate cover story using fake bank accounts and government contracts. And when curious reporters started calling, Hughes agreed to field questions about the project with evasive non-answers.

The cover story held up perfectly, and the ship was designed and built in secrecy. The *Glomar* included a derrick similar to what would be found on an oil-drilling rig. It had a crane that extended by adding or subtracting pipes, along with two tall docking legs, a claw-like capture vehicle capable of holding the entire submarine, a center well for docking the vessel once it surfaced, and payload doors that would open and close on the bottom of the rig. When finished, the James Bond–like vessel with hidden compartments and secret capabilities could conduct the entire operation *underwater* without being seen by other ships, aircraft, or spy satellites. While ocean currents jostled the platform, the ship had to lower the capture craft by adding sixty-foot sections of pipe one at a time. Once it reached the Soviet sub, the vehicle would hover above the wreck and close its jaws. Then the pipe sections would be removed one at a time until the Russian sub arrived in the ship's docking bay.

Like Woolley's plan, the difference between theoretically possible and technically viable soon became clear. After six years of preparation, the *Glomar* arrived at the wreck site on July 4, 1974. The first mishap occurred several days later, when the crew spotted a missile-tracking Soviet ship approaching. The Russian ship, the *Chazhma*, radioed the *Glomar*.

"What are you doing here?" the Russians asked.

The *Glomar* responded, "We are conducting mining tests—deep-ocean mining tests."

"What kind of vessel are you?"

"Deep-ocean mining vessel," the Americans said.

"How long will you be here?"

"Two or three weeks?"

Five minutes of tense silence passed. Everyone on board the *Glomar* wondered if their cover had been blown. Further interrogation might have revealed the true nature of the equipment on board, or, worse, being watched would have nixed the chance to activate the *Glomar*'s intricate underbelly.

Finally the *Chazhma*, largely appeased and low on provisions, radioed back, "We wish you all the best."

Several days later, the descent went as planned. The crew watched on closed-circuit television and marveled at the richly detailed crabs and fish they could see around the wreck. The anchoring and clamping of the claw proceeded smoothly, surpassing the most difficult phase of the mission.

During the lift, however, when the submarine was a third of the way up and had ascended more than a mile, the video feed went black. When it was restored, the camera that had previously been pointed at the submarine now showed nothing but water. The front thirty-eight feet of the submarine was still in the claw, but the rear one hundred feet had fallen back to the seafloor. The lift continued on the front half, which arrived minutes later on the surface. It contained the bodies of six Soviet crew members (who were subsequently buried at sea in metal coffins) and two nuclear torpedoes, code books, sonar equipment, and navigational documents.

The mixed success of the operation might have spurred a second attempt to recover the fallen section, if not for an odd turn of events. While the *Glomar* and its crew planned a return mission, burglars with no knowledge or apparent interest in the *Glomar* broke into the headquarters of one of the shipping contractors in Los Angeles and stole documents, including one tying Howard Hughes to the CIA. The FBI was assigned to recover the document, which later turned out not to be among the stolen documents and was found by a security guard. But it was too late. Word spread to several investigative journalists, who broke the story in the *Los Angeles Times* on February 8, 1975. The Nixon administration denied the story, but the Soviets weren't swayed, and they deployed a ship to sit atop the recovery site. William Colby, the director of Central Intelligence, tried to suppress further reporting on account of national security against America's greatest foe. But a well-known muckraking radio journalist named Jack Anderson wasn't convinced. Navy sources told Anderson that the operation yielded no actionable secrets and was mostly a waste. Later, on his radio show, Anderson called the billion-dollar operation a government "boondoggle" and said taxpayers deserved to know about it.

What would it take for a government-sponsored salvage operation to be supported by the general public? America got its answer a decade later in a disaster eerily similar to the *Titanic*'s. In a parallel tragedy that involved overextended new technology, the death of a cross section of humanity, and the disastrous effects of ice, the *Challenger* space shuttle exploded one minute after its launch over Florida in 1986 and fell, like the *Titanic*, to the floor of the Atlantic.

By the time the first replays of the explosion were airing on television, NASA already had a plan to salvage the components of the shuttle. And a week later, the U.S. military and NASA commenced a four-month-long, around-the-clock search of a four-hundred-twenty-square-mile area. Urgency was crucial before debris could float away or be buried by mud. If found, the spacecraft structure and payload components would help ascertain the cause of the explosion. Recovering the hazardous components would also prevent fuel, oil, or other chemicals from polluting the ocean. Most morbid of all, finding the crew compartment would be the only guarantee that mangled limbs of astronauts didn't wash up on Florida beaches and make matters worse.

The *Challenger* explosion was a dramatic American moment, but like any disaster, its emotional punch was in its explosion, not its cleanup. I heard about the salvage operation for the first time from a handyman who was caulking the windows of my house one day. He mentioned he was a former deep-sea diver, and I told him I was working on a book about the deep sea. He suggested I talk to his friend who worked on the *Challenger* recovery in 1986.

"Recovery?" I said. "What recovery?"

"We basically had to go pick up all the parts," his friend, Paul Kowalski, told me when I called him.

Kowalski was a navy diver stationed in Washington, D.C., and was one of six thousand pairs of hands assigned to all be on deck, more than half of them divers. The navy provided an NR-1 nuclear submarine along with sonar rigs, remote-operated vehicles (ROVs), manned submersibles, and metal detectors. They called in a ragtag team of deep-sea professionals known to gather for high-profile sea hunts, including for wreckage of a Korean Air Lines plane shot down by the Soviets in 1983 and an Air India jet that exploded near Ireland in 1985.

Once the sonar hit on something, Kowalski and other divers were sent down to check it out. Only one in eight hits were actually shuttle-related. The rest were oil drums, forgotten shipwrecks, and even an old DC-3 airplane that crashed in the 1930s. Sonar is notoriously bad for revealing shapes, and it fell to the "gut feel" of a few deep-sea veterans to tell if an object was a shuttle part or simply a fish hole or depression in the sand. Kowalski hauled up pieces of wreckage if he could or directed an ROV to the site. No one knew what happened to the debris after that, but rumors went around that NASA buried them in concrete bunkers to put them—and any radioactive components—out of sight and out of museums. The last thing NASA wanted in the Smithsonian was a memento of the agency's worst day.

In the end, the search was fruitful. The salvage operation recovered fifteen tons of debris, including half of the orbiter and more than half of the rockets. Six weeks passed before an unmanned vehicle discovered a window frame and a blue cushion, signs of the mangled crew cabin sitting in one hundred feet of water about seventeen miles off shore. The bodies were removed one by one—except one of them, payload specialist Greg Jarvis's corpse, which floated out of the compartment during the rescue operation and was lost by the divers before they recovered it weeks later on the seabed. The search confirmed that the cause of the crash was weak O-rings unsuited to freezing temperatures that separated the fuel segments, a finding that helped NASA reengineer the space shuttle program. By the end, the entire operation had cost $13 million—a hefty but affordable sum to pick up the pieces of a national tragedy.

As luck would have it, on April 12, 1986, one month into the search for the *Challenger*, sailors noticed a piece of debris floating near the wreck site that wasn't related to the shuttle at all. The object turned out to be a duffel bag filled with a dozen bricks of high-quality cocaine that one could only assume fell off a smuggler's boat. To the extent anything about the *Challenger* was even remotely funny, the navy crewmen joked that the salvage effort wound up paying for itself. By coincidence, the drugs were valued at exactly $13 million.

° ° °

Looking at a shipwreck underwater is usually good enough. The vast majority of wrecks ever found have been gawked at by divers breathing from oxygen tanks. *Lifting* a boat, however, is accompanied by the requirement to do something with it, which frequently costs more than the salvage itself. In 2002, archaeologists found a fifteenth-century sailing ship near the town of Newport in Wales. Local people got interested in restoring the ship, which they nicknamed the *Newport Ship*, and tried to fundraise £3 million. But costs swelled when accounting for keeping an old ship in presentable shape in perpetuity.

On a larger scale, the company that owned the *Costa Concordia*, the cruise ship that famously wrecked off Tuscany in 2013, debated for about five minutes whether to restore the ship to its pre-capsize grandeur. They decided that the more than €1 billion salvage operation was, in every sense, a sunk cost and sent the wreck to be scrapped.

The complexity of salvage can also make it painfully boring. Like building an amusement park or passing a law, the process is far less interesting than the finished product. In 1980, the film *Raise the Titanic*, based on Clive Cussler's classic novel, was a commercial flop because the title was the most breathtaking part. The plot followed a race between the U.S. and Russia to acquire a powerful made-up element called byzantium that held the potential for global dominance and yet only existed in a safe aboard the sunken *Titanic*. The fictional premise made for a decent book, but the film was such a box-office dud that it ended up losing more than $30 million. "Raise the *Titanic*?" one critic wrote. "It would have been cheaper to lower the Atlantic." The plot was mired in technical details about metallurgy and fluid dynamics that made the film, according to another reviewer, "as creaking and slow-moving as a battleship attempting a U-turn." Few in Hollywood have since tried to make a blockbuster film about the intricacies of deep-sea engineering.

By January 1970, Doug Woolley began promising that *his Titanic*, the real one, would be out of the water within a year. That alone turned heads and yielded money. Woolley was often tight-lipped about how much he raised and from whom. But he announced on January 12 that a firm in West Germany had promised £10,000 and that, if all went as planned, the team would arrive at the wreck site by April. By March, little progress had been made on the preparations. But the team had expanded to now include a doctor, a nurse, two electronics experts, an Australian pensioner, and a twelve-year-old schoolboy, who Woolley suggested should be the team's "mascot." Using his connections in textile manufacturing, Woolley also convinced a Scottish clothier to donate a dozen coats for the crew heading to the Atlantic.

"So far, everything is proceeding absolutely on schedule,"

Woolley said in April. He claimed to have £10,000, a sum he believed sufficient for a preliminary survey to visit the patch of ocean where the ship was thought to have sunk. April came and went with no visit to the site. Woolley claimed he was still talking with experts and raising money, but he also began to distance himself from actually doing the complex work of deep-sea salvage. "We are confident," he told the *Reading Evening Post*, "that practically every phase of the operation—the lifting and the special equipment—will be tackled by professional companies," which he believed would donate their time and equipment in return for the "prestige and publicity" of participating. Two months later, Woolley pushed back the schedule another three months. A year later, he pushed it back again. All the while, his name continued to appear in the papers as the man on the verge of doing the impossible.

How could journalists buy such a technical scheme dreamed up by such an unqualified enthusiast? Why didn't anyone hold a magnifying glass to expose such an obvious house of cards? It's hard to pathologize British reporters of the 1960s, but I'm sheepish to admit that in my experience, it's easier than it seems to get caught up in a news story that's too exciting not to be true. Many years ago, I proposed a story to my bosses at *National Geographic* about a group of young botanists in London growing lettuce in old World War II bomb shelters. The story had strong themes—past meets future, wartime tragedy fuels modern innovation, urban concrete becomes clean and green—plus the chance for great photos. Even though it was my job to be skeptical, it was also my job to tell colorful stories with vivid characters, and this story seemed to have it all. My editor enthusiastically agreed, and the next time I was in London, I scheduled an afternoon with a photographer for us to visit the tunnels. When we got there, it became clear that a dozen other journalists and TV camera guys had also been taken in by the story. We all marched down to the bunker and scribbled on our notepads as the two men made grand visionary statements to describe their brilliance. We went home and published the story, which became one of the most popular stories I ever wrote. Thousands of people read it online; many commented and sent me excited emails about it. Only years later, when I realized I hadn't heard a single follow-up word about this supposedly transformative idea, did it occur to me that the whole thing was an empty news gimmick, a story without a center. We all were too wrapped up in an enchanting tale to consider what now seems obvious: that energy-guzzling farms in dank concrete tunnels could hardly feed a single person, let alone the world. Reality came a distant second to the chance to tell a gripping tale, and we bought it wholesale.

You can fool reporters only so many times on the same story, though. And in late 1971, Woolley's facade started to show cracks. The money from the West Germans never came. The Hungarian craft was never quite ready, or maybe it didn't exist at all; Woolley had never seen it. He wrote to several salvage firms that specialized in deep-sea maneuvers, and all either laughed at the question or quoted him an absurdly high price to lead an operation they knew had little chance of success.

But these were mere technical matters in a sea of existential concerns. By now the project had firmly entwined itself with Woolley's personal identity. The *Liverpool Echo* reported in late 1971 that Woolley "staked his existence on lugging the great ship back to life." Rescuing the *Titanic* was his "lifelong ambition," another profiler wrote. Woolley, a bachelor at thirty-eight, told friends that "The Big T is my first and last love."

Under such immense expectations, a weaker man might let the

whole house of cards fall and move on. But Woolley had invested too much. The only way he'd let go of the *Titanic* was if the frigid ocean itself pried it out of his cold hands. If he had to, he was prepared to go down with the ship.

ooo

Would it have made a difference if Woolley and the rest of the world knew where the *Titanic* actually was? Would it have been easier if the search *and* recovery mission was minimized to recovery only? Needles in haystacks don't often identify themselves, but yes, a precise location would likely have downgraded the difficulty of lifting the *Titanic* from preposterous to merely absurd.

Wrecks far closer to shore in much shallower water remained mysteriously hidden from even advanced wreck hunters for decades longer than the *Titanic*. The USS *Monitor*, the famous ironclad Civil War battleship that sank two hundred fifty feet deep off North Carolina in 1862, evaded investigators from the U.S. government and the National Science Foundation for more than a century, until 1973, the same year Woolley was running out of steam, when the old steamship appeared off the coast of North Carolina on recordings from a sonar rig towed behind a research boat. It took three different cameras to confirm it was the *Monitor*. Pulling it out of the water didn't begin until the Clinton administration.

For almost all of history, humans who roamed the planet on boats believed that boats sunk were almost always boats gone. But with the turn of the twentieth century came a soaring interest in the natural world. One can picture a sprightly Teddy Roosevelt leading this new growth of oceanic awareness, but in fact it was Prince Albert of Monaco, the ocean buff formerly obsessed with the Gulf Stream. In 1910,

Albert commissioned four research yachts to tow sonic and ultrasonic devices around the Mediterranean. The data produced the first underwater maps a year after the *Titanic* sank. It would be decades before sonar could capture the rocky and muddy nuances of the seabed's endless texture—and even longer (we're still waiting) before this technology could be deployed worldwide to map the entire seafloor.

A century later, there's a common line repeated by marine scientists: that we know more about the surface of Mars than we do about the ocean floor. By and large, it's true, because Mars is easier to map. Land-based maps are the result of orbiting satellites emitting micro radio waves. But radio waves get stuck in water and don't bounce back. So the easiest way to map the bottom of the ocean is in fact to map the *surface* of the ocean, which is never truly flat. Underwater gravity pulls and pushes water in strange ways. An undersea mountain five hundred feet tall will raise the sea surface above it about one foot. The bumps or depressions are even more dramatic around the ocean's major geologic features, like the Mariana Trench or the Mid-Atlantic Ridge. In 2014, a group of scientists discovered that undersea topography and swirling currents have the effect of *sloping* the oceans. They found that the sea level off the coast of Europe is three inches lower than off the East Coast of the United States.

More detailed elevations are impossible to see from space; in fact, they're visible only using expensive and clunky sonar rigs towed behind ships that send sound waves down and convert the time of their bounce back to an estimated distance. This explains why every high school science classroom has a map showing the broad contours of the seafloor, but not even the best scientists on earth using the most modern tools can find history's most famous undiscovered shipwrecks, like Columbus's *Santa Maria* or the SS *Waratah*, a 1909 passenger ship nicknamed Australia's *Titanic*. More modern searches are just as vexing. In 2014, after Malaysian Airlines flight 370 disappeared over a region of the Indian Ocean known for extreme depths and punishing surface conditions, three countries spent three years combing nearly fifty thousand square miles of the seafloor using the most advanced sonar technology in existence for any sign of the airliner or the 239 people on board. They abandoned the search without finding a single trace.

If Doug Woolley had better eyes underwater, he would've had a head start. He could have been the person who *found* the *Titanic* before he became the person to raise it. In his mind they were synonymous: Why do one without the other? But the world was hungry for news on either front, and he was beginning to feel the pressure of a crowded field. If he continued blowing past deadlines and dithering with a team of amateurs, someone smarter, wealthier, and better equipped would beat him to it.

There was another way, he realized. Not to commence operations on the *Titanic*, but to buy time and prove his concept. And in the process, to stage a dress rehearsal that would shore up his self-proclaimed expertise and silence the doubters. If he could raise one ship—any ship—using his method of pontoons and hydrogen, then the world might believe he could raise a more difficult one.

But which ship? A letter from a stranger stuffed with a newspaper clipping caught Woolley's eye in the spring of 1972, and within two months, he was flying to Hong Kong for an audience with Queen Elizabeth herself.

Chapter 8

# TAKE ALL THE BODIES AND TREAT THEM WITH RESPECT

Queen Elizabeth traveled to Clydebank, Scotland, on September 27, 1938, the day the RMS *Queen Elizabeth* was to be christened. The ship had been built on a dry dock next to the river Clyde at a slight angle, so that once completed, it could slide on wooden rollers down the gangway and hit the water with minimal splash. Thirty thousand people arrived when the queen consort, the woman who was married to King George VI and would later be known as the Queen Mother, arrived to coronate the eighty-three-thousand-ton ship named for her. She also brought her children, one of whom, also named Elizabeth, would become the next, long-standing monarch.

The *Queen Elizabeth* was a ship built for royalty, bigger and better than any ship that came before. It was designed particularly to eclipse its sister ship the *Queen Mary*, which had by then been in operation just two years. The *Elizabeth* was bigger than the *Mary* by eleven feet and four thousand tons, but significantly bigger than the *Titanic* by almost 50 percent. In a time darkened by the prospect of war, the queen used the occasion to try to fortify the resolve of the British people. "The launching of a ship is like the inception of all great human enterprises, an act of faith," the queen said, the first time the general public heard her voice carried over the airwaves. "We cannot foretell the future, but in preparing for it, we show our trust in a divine providence and in ourselves."

When it came time to launch the ship, an engineer on the ramp moved a supporting beam before he was given the cue, and the ship began to slip down the ramp while the queen was still shaking hands. Acting quickly before the *Elizabeth* rolled away, the human Elizabeth grabbed a bottle of Australian wine and smashed it against the hull. Few noticed the hiccup. Everyone cheered, and the band played "Rule, Britannia," the classic hymn of British patriotism.

That was the *Queen Elizabeth*'s best day. Its worst came thirty-four years later on the other side of the world. The grand ship lived an exciting life. During the war, it carried troops to battle as one of the safest and fastest boats on the water that could outrun German U-boats. After the war, the *Elizabeth* was retrofitted to its original purpose as the grandest ship on the Atlantic, and owing to its storied past and glamorous present, it attracted celebrities, politicians, and royalty.

The *Titanic* may have taken the same path had it not sailed a generation too early and met its fate too soon. But all ships eventually meet the great shipmaker in the sky, and on January 9, 1972, after the *Elizabeth* had been sold to a series of buyers for lower and lower prices, and after a Chinese man bought the ship intending to turn it into a floating university, the ship caught fire in the harbor of Hong Kong while undergoing renovations. Firefighters battled the blaze for two days, but the water used to fight the fire caused the *Elizabeth* to list and capsize.

Doug Woolley was in his apartment in Baldock on January 11, 1972, when he read about the *Queen Elizabeth*. He had followed the ship's entire trajectory, watching its transformation from warship to passenger liner, following it from Scotland to California and around the Cape of Good Hope. A man who made barely £50 a week could only dream of sailing on the jewel of the British shipping industry, but that didn't stop him from collecting memorabilia and newspaper clippings from the *Queen Elizabeth*'s illustrious life on the high seas, including used tickets, a captain's hat, and a small-scale replica. To shipping buffs around the world, the *Elizabeth*'s smoky demise sparked the conspiracy theories that always follow a major tragedy. But Woolley was more enchanted by another component of the *QE1*, a nickname that became necessary after the *Queen Elizabeth 2* was christened by Queen Elizabeth II in 1967. The freshly sunken *QE1* solved one of the most vexing challenges of Woolley's plan for the *Titanic*: depth. The *QE1* had sunk in shallower water than had the *Titanic*, making it far easier to test the hydrogen-pontoon method, prove its success, raise money, and build new momentum.

The first step was certain because he had done it before. He would go about obtaining salvage rights to the *QE1*, fly to Hong Kong, and finally put his decades of calculations and enthusiastic teammates into the water and show the world that a man who had lived his whole life doubted and mocked was about to pull up a sunken ship.

° ° °

Woolley arrived in Hong Kong in late April 1972 after the longest flight of his life. It was only the second flight he'd ever taken. The first had been to Hungary several years earlier to meet letter writers who begged to join his team after they read about his plans for the *Titanic.* His travels on airplanes enchanted him. Not only was he traveling internationally, he was traveling on *business*, a privilege rarely afforded factory workers. Between plates of goulash and pints of beer, Woolley reveled in the way he was moving and shaking as an international entrepreneur. "They treated me better than anyone treated me in my life," Woolley said.

The trip was a high-stakes gamble. He had quit his job at the pantyhose factory and let go of his apartment in Baldock, not knowing how long he'd be in Hong Kong to get the *QE1* out of the water. He sold his furniture, put a few valuables in storage, and packed everything else he owned in a small suitcase.

Traveling to Hong Kong raised the stakes. Several weeks before he left, Woolley began to feel like he was being followed. News of the tragedy of the *QE1* was quickly overtaken by news of the fire, and specifically that it may have been arson. An investigation revealed that the blaze had started in nine different parts of the ship at the same time and was fueled by flammable materials not often sitting idle on ships. Attention swung to the last owner of the *QE1*, the Hong Kong businessman Tung Chao-yung, who in its final days afloat changed the ship's name from *Queen Elizabeth* to *Seawise University*, a campus of the World Campus Afloat program, which would later become Semester at Sea. No one was able to prove the fire was Tung's idea, but Hong Kong reporters pointed out the coincidence that the ship's renovation had significantly overrun its budget and that the vessel was insured for twice its assumed value.

If Woolley was reluctant to get caught in this web of allegations and crime, he acted the opposite. The same way he spread rumors about his ownership of the *Titanic*, he began to mention his desire to lift the *QE1*. He changed the name of his salvage company from *Titanic* Salvage Company to Seawise Salvage to associate himself with the name of the ship. Predictably, the newspapers followed suit and began to report that Woolley was aligned with the *QE1*, which wasn't entirely true, and that he had bought the salvage rights of the *QE1* for £3,000, which wasn't true at all.

Before he flew to Hong Kong, Woolley's sleight of hand caught up with him one night while he ate greasy fish and chips with friends at the local chip shop in Baldock. The group made the spontaneous decision to go to one of the friend's houses in Liverpool, and once they got there, the phone rang. The caller asked for Woolley and informed him that some businessmen on the island of Jersey needed to meet with him immediately. Woolley told this story years later in his biography of himself, which he wrote in the third person. "They told Doug to go to Southampton and get on the plane that was waiting for him at the airport to take him to Jersey, as it was of the utmost importance that they signed a contract as soon as possible," he wrote. The businessmen, in Woolley's telling, worked for another salvage company and wanted Woolley to relinquish his supposed rights to the *QE1*. Under the promise that the businessmen and Woolley would work together to raise the *QE1*, the contracts were signed and backs were slapped. "Doug was very bemused by the whole thing and did not know quite what to think," Woolley wrote of himself. He returned home excited at the promise of partnerships and progress. He never heard from the businessmen again, but he later suspected that they were working for Tung, the former owner of the *QE1*, who wanted Woolley sidelined to prevent him from pulling up evidence that the fire was intentional.

Whether this indeed happened is a mystery lost to time. Woolley's repeated claims that he would raise the *QE1* as a dress rehearsal for the *Titanic* were widely reported. When questioned about his stake in either of the ships, Woolley would point to prior news clippings that confirmed his involvement, which often satisfied the asker. Court filings were never produced; judges were never petitioned. From the moment the *QE1*'s fire broke out through the following years of legal squabbles, salvage attempts, and jockeying for any valuables on board, the title for the *QE1* lay in the most obvious place: with the underwriters who insured it back in London.

Perhaps predictably, Woolley left Hong Kong in the summer of 1973 with nothing but fond memories of Asia. Despite the easier challenge of raising a ship one hundred twenty feet rather than twelve thousand five hundred feet, Woolley was genuinely flummoxed that the British and Chinese officials who governed the waters, and Lloyd's of London, which governed the ship, didn't entrust him, with his lengthy résumé of press clippings, to undertake a high-stakes project of engineering next to one of the world's most populous cities.

While Woolley wandered the streets of Hong Kong on his final night in the city, a government commission made the decision that the fire rendered the *QE1* in too poor condition to be refloated and repaired. The only option was to remove a once great ship by amputation, one piece at a time.

∘ ∘ ∘

The collapse in Hong Kong dealt a heavy blow to Woolley. He hadn't failed at raising the *Titanic*; he had failed at a project far easier than raising the *Titanic*, which didn't bode well or look good. Like anyone who markets in public enthusiasm, Woolley knew his lack of success could be explained away in the short term, but eventually, the wave that carried him to the center of attention would break and whitewash him over.

Woolley arrived back in London with little more than the clothes he was wearing. On the journey back from Hong Kong, his suitcase was lost, a mishap that fed his suspicion that his international reputation and his widely reported salvaging plans had made him enemies.

In the story of Doug Woolley, this is his rock bottom. With few close friends, no money, no home, and no job, Woolley became homeless. From the airport, he found his way into London and for two weeks slept on a park bench in Chelsea beneath a statue of St. Thomas More, a Catholic saint executed for swimming against the changing tide of the Protestant Reformation. He subsided on donations and ate the expired items thrown away from nearby markets.

One night a man walking through the park asked Woolley why he was sleeping on a bench.

"I've just returned from Hong Kong, where we did a survey on the *Queen Elizabeth*," Woolley told him.

Woolley explained his work and that he owned the salvage rights to the *Titanic* and, by extension, he practically owned the *Titanic* itself.

The man happened to work for the BBC, and when he went to the library and confirmed Woolley's claims in newspaper write-ups, he invited Woolley to stay with him and then found him a small apartment. All his life Woolley obsessed over saving the *Titanic*, putting his entire life's energy into rescuing a piece of faraway, mangled metal. And yet, when his most dire moment arrived, it was the *Titanic* that saved him. With the BBC fellow's recommendation, Woolley got a job in a North London factory that made machinery parts. He found a room in a boardinghouse and, also with the BBC man's encouragement, wrote to his member of Parliament, who urged the local council to find him a flat of his own. Using nothing but his association with the *Titanic*, Woolley in six months went from destitute and jobless to nicely employed and living in a large flat in the North London neighborhood of Edmonton. He thought the best part about his new home was its blank walls, which he hoped to fill with future photos and news clippings.

Woolley still had muscle memory for building hype. He returned to planting stories in the papers and recruiting supporters, often in unlikely places. One night while walking through Edmonton, he noticed two feet sticking out from under a car. He checked to make sure the man was alive, and then invited him inside for tea. The young man, named Steven, said he was an artist and was taking an evening course in deep-sea diving. Woolley offered him a sleeping bag and his couch, and a week later, he made Steven second-in-command of his company, which he renamed for the third time, *Titanic* and Seawise Salvage.

Meanwhile, there was another setback brewing. Two decades of breakneck innovation in undersea exploration, treasure hunting, and deep-sea mining had given way to a new area of research that directly objected to the notion of pilfering the ocean for valuables. A discipline called maritime archaeology was a new arm of conventional archaeology, where brush-holding scientists unearthed dusty pots and tiles. Performing this work underwater was newly possible in the 1970s with advanced diving tools and waterproof equipment that allowed a diver to spend hours and sometimes days submerged. Growing interest in maritime history signified a cultural change that old wrecks—and especially famous ones—weren't to be exploited, manipulated, and least of all salvaged, but rather should be preserved and studied.

George Bass, a Texas scientist, was the first person to call himself an underwater archaeologist. Bass started in the sixties looking for ancient wrecks in the waters off southwestern Turkey, a ship graveyard owing to its rocky outcrops and strong winds. When he found Byzantine wrecks from the fourth and seventh centuries and pulled up old lamps and remnants of ancient silk, he won the kind of attention that Woolley craved. Scientists flocked to the area. Newspapers trumpeted the findings. Bass was invited to speak at universities.

Bass was a prolific diver. Every new discovery he made built the public case for appreciating wrecks with caution, not abandon. The best case of this was a two-masted Byzantine vessel filled with broken glass that Bass dated to the eleventh century. "We must have raised 1,000 amphoras that all looked identical," Bass said. "But then one of the Turkish graduate students noticed graffiti on the glass, and that graffiti alone enabled us to determine that the ship belonged to a church and was carrying wine over land and sea to Byzantine troops in a certain city." Bass began testifying in court against treasure hunters, who he believed were ransacking history's priceless artifacts.

Woolley wasn't a treasure hunter. But he wasn't a soft-handed scientist, either. His plan to lift the *Titanic* was in direct conflict with a growing sense of archaeological ethics that urged caution before upturning items that couldn't be put back. In addition to the diaries, paintings, and wood carvings that Woolley wanted to bring up from the ship, there was also the matter that the *Titanic* was a mass grave. Responding to calls that the wreck and its contents—including human remains—be recovered for the sake of science, survivor Eva Hart said the victims had been through enough. "The ship is its own memorial," she said. "Leave it there."

Plenty of people disagreed with such wrought oversentimentalization, including Woolley. Seventy years was long enough for bodies to rest in peace, if any still existed at all. "Take the archeologists in Egypt, for example," he told an interviewer. "The pyramids are broken into and the Egyptian Mummies are exhumed for all to see." Wouldn't it be better if the remnants of history's most famous shipwreck were in a museum, rather than hidden away? Wouldn't the risk of jostling a historic site be worth the educational effort of being able to study it up close? It didn't take much for Woolley to convince himself that jostling a grave site was actually the prudent thing to do. In the days after the *Titanic* sank, more than three hundred floating corpses were found and buried, many of them in a mass grave in Nova Scotia. If a recovery mission could raise more body parts, they too could be buried in a cemetery on land, where descendants could visit them. "We will take all the bodies and treat them with respect," Woolley said.

The argument had merit. But those who agreed with him were also his competition. In 1975, the American-based Titanic Historical Society had ballooned to a worldwide membership of one thousand. All meetings of the organization and editions of its newsletter, *The Titanic Commutator*, dissected the endless details and inconsistencies of the sinking. But the biggest question hanging over the group was when the wreck would be found.

The natural inclination for any group of wreck enthusiasts is to go out and look for one. But Edward Kamuda, the group's founder, knew his membership was better suited to armchair expeditions than real ones. Unlike Woolley, who insisted all decisions and operations went through him, Kamuda believed scientists would be the first to find the *Titanic*. He thought his group's best shot at credit, glory, and proximity to famous wrecks like the *Titanic* was to open the group's voluminous archive of historical data and expertise to credible marine scientists to search for clues. Such an arrangement of openly sharing information is extremely uncommon in wreck hunting, especially without fine-print contracts.

But it worked. In 1975, Kamuda's group heard that the famous ocean scientist Jacques Cousteau was in the Aegean Sea off Greece looking for the lost city of Atlantis. Kamuda and several THS officials contacted Cousteau and suggested that, while he was there, he should look for the *Titanic*'s sister ship, the *Britannic*. The *Britannic* had sunk off the coast of Greece in 1916 after being struck by a German mine, but in the flurry of war, it was never found. Cousteau liked the idea and asked the group for more information.

Mustering every detail it could find, the THS sent a lengthy dossier to Cousteau about the *Britannic* and the circumstances of its sinking, along with its final emergency coordinates. Cousteau reviewed the documents, and in November 1975, he directed his ship, the *Calypso*, to the probable site. For several weeks he scanned the seabed with a side-scan sonar vessel. Finding nothing, he slowly expanded the search area, and a month later he came upon a wreck more than six hundred fifty feet long buried three hundred feet deep. The *Calypso* circled the site with the sonar rig in tow while Cousteau inspected the damage from the mine. A year later, he swam through the crew quarters, which appeared to be in excellent shape.

Word of a newly discovered 1910s-era passenger ship and the revelation that it was well preserved boded extremely well for future efforts to recover the *Titanic*. But the success came at a cost for Woolley. In his quest to go it alone and build his own team, he had declined to join the THS and ride the coattails of the *Britannic*'s triumph. He had declined to share any of his ideas with scientists like Cousteau who had the skills and equipment to explore underwater. And he had failed in Hong Kong to prove his ability to raise a wreck himself. His ambitions had largely been eclipsed by an alternate approach fueled more by expertise than emotion. When the Associated Press reported in July 1976, "Cousteau Strikes Gold Below Sea," it was as though they'd stolen the headline Woolley craved most and given it to someone else.

∘ ∘ ∘

It took two years for Woolley to climb back after his Hong Kong disappointment. Having jumped from job to job for most of his life, he lacked the skills and résumé that might have earned a man of his age higher-paying work. His obsession with the *Titanic* had cost him old friendships and seemed to turn off women.

When he had a few hundred pounds saved, he moved to the East London neighborhood of Ilford, where most of his neighbors were immigrants from Pakistan and India. He furnished the apartment first with his albums of press clippings and then with any memorabilia he could find—ship replicas, old cargo certificates, peaked captain hats he picked up at thrift stores. Anything with the name *Queen Elizabeth* or *Titanic* he regarded as gold.

For a while he continued his earlier approach, drumming up attention and leveraging it to earn more in hopes of building enough interest to raise actual money. But the strategy brought diminishing returns. Following the revival of the *Titanic* in Walter Lord's book and the film *A Night to Remember* in the 1950s, the *Titanic* in the seventies had begun to recede from public view the same way it faded in the years after it sank. In Woolley's early days, he could drum up enthusiasm with a well-placed phone call. Now, much more consequential news had overtaken the kind of frivolous reporting that had once gotten Woolley so much ink. America was at war in Vietnam. Miners and other wageworkers were repeatedly on strike in England. Clashes between Protestants and Catholics had erupted into bloody killings that occupied the British military, the economy, and people's idle energies. The idea of reviving an aging shipwreck that predated most people alive didn't make the cut.

Adding to the *Titanic*'s receding relevance, finding old shipwrecks had become easier. Nearly every month in the early seventies saw the announcement of a new shipwreck discovered. In July 1973, divers came upon more than a hundred wrecks in the Ottawa River in eastern Canada. Farther south, in Texas, the state antiquities commission sponsored a search for sixteenth-century Spanish wrecks in the waters off Corpus Christi. Two years later, a seventeenth-century Spanish galleon turned up in South Carolina. And by the end of that summer, an international search of ancient trading routes led by *National Geographic* announced the find of a lifetime: a Cycladic-era trading vessel dating back as many as 4,500 years.

Money was required to sponsor these expeditions, and generally, well-funded searches turned up big discoveries. A large fishing trawler sank near Cape Cod in February 1978, and two weeks later, divers found it and recovered the bodies of four crew members. Farther north, a well-funded Canadian archaeologist happened upon the *San Juan*, a Basque whaling vessel lost in the icy waters of Labrador in 1565. A year later, a privileged group of international students on an around-the-world voyage called Operation Drake discovered almost by accident a sunken 1699 trading ship on the Caribbean coast of Panama. The announcements began to blend together—did people really care if a wreck was from the sixteenth century or the seventeenth? As the general public grew numb to the sheer repetition of shipwreck discoveries, the announcements dropped from page A-1 to the inner sections.

Yet for those motivated by money, the more lucrative deep-sea searches bypassed shipwrecks completely. "On the bottom of the ocean are mineral deposits large enough to supply all mankind for years, even centuries, to come," reported *The New York Times* on July 17, 1977. "It's a hunt for sunken treasure on a corporate, national, and global scale." Experts imagined six thousand years' worth of copper, twenty thousand years' of aluminum, and one hundred fifty thousand years' of nickel sitting in rocky deposits under the seafloor. The estimates didn't account for a future of personal electronics that metabolized rare earth elements much faster than anything in existence in the seventies, but the quantities underwater were (and still are) eye-popping.

Mining introduced another species of deep-sea hunter into a field already crowded with scientists and shipwreck hunters. The oceans, which once seemed empty and limitless started to feel small and crowded. In 1972, an ocean ecologist named Hjalmar Thiel went to a part of the Pacific Ocean known as the Clarion–Clipperton Zone, an area full of both new biodiversity unknown to science *and* rare-earth element deposits of copper, nickel, and manganese. To save money on the trip, Thiel split the ship cost with prospective miners, who made their intentions known in advance. "We had a lot of fights," Thiel said after the trip. He told the miners that if they dumped their dredged-up sediment, it would smother plankton and domino up the food chain. The men resented the buzzkill. "They were nearly ready to drown me."

The ability to drill into the seafloor required a new technology that enabled a ship in deep water to position itself directly above its target and stay in that exact same place without an anchor for hours, sometimes days. Wind and water currents made this impossible until 1961, when an engineer for Shell named Howard Shatto realized that if a captain could maneuver a ship front to back *and* side to side, he could conceivably counteract any motion long enough to pull up a geologic sediment core. For most boats an anchor will suffice, but in high-stakes and high-cost situations, staying perfectly still comes with no room for error. A decade later, Shatto's system, called "dynamic positioning," was co-opted and became standard on all oil-drilling ships.

This solved the surface issue, but still elusive was the ability to lower a drill several thousand feet into igneous rock below the seabed. Often the drill bits would wear down before they reached the desired depth, which required the crew to extract the spent bit, replace it, and guide it back through several miles of water into the exact same hole. One observer compared it to lowering a strand of spaghetti into the drain of an Olympic swimming pool. The first time it was done successfully was in 1970.

Oil was the first undersea commodity worth getting. But the new tools capable of drilling faster and deeper were effective only if you knew where to drill. The oil companies with the most to gain, ones like Shell and Schlumberger, developed software to inspect a piece of seabed and instantly calculate advanced metrics like fluid saturation and multimineral lithology. Advanced vocabulary has always been the tool of scientists, but in the case of mega-profit oil exploration, technical jargon helped mask the ugly reality that loosely regulated corporations were cracking open the seafloor to pull up toxic sludge that, in even minor accidents, could devastate marine life.

The extent of this danger became clear in 1969, when a blown-out well off Santa Barbara, California, ruptured the seabed in five places and spilled three million gallons of oil in one of the world's most biodiverse coastal ecosystems. A year later, twelve wells off the Louisiana coast exploded into a fire so hot and powerful it burned for four months. Such horrifying accidents spawned an environmental movement that led to the creation of the Environmental Protection Agency and a host of laws regulating everything from the use of pesticides to the limits of oil excavation on the seafloor. But drilling rigs still grew bigger and their depths deeper, along with a class of ships of their own—oil barges, submersibles, platforms, floaters, and jack-ups. One of the biggest drilling rigs ever built, a semi-submersible called the *Deepwater Horizon* and operated by British Petroleum, blew up in the Gulf of Mexico in 2010. For five months, an uncapped well spilled more than two hundred million gallons of crude oil in the temperate waters of the Caribbean. Before the platform exploded and sank to the bottom of the Gulf, it had been capable of drilling thirty-five thousand feet deep, almost seven miles below the sea surface and three times deeper than the wreck of the *Titanic.*

One could imagine the tremendous boost this technology could give Doug Woolley. Rather than bob aimlessly with a team of amateurs above where they thought the wreck might be, the ship that could hypothetically pull up the *Titanic* could stay perfectly still above the precise wreck site. It could send down a drill to barrel deep into the mud that would eventually give way to rock. Then it could drive long steel pipes into the holes and build an entire scaffolding around the wreck. The same remote-operated vehicle that guided a hydraulic drill could bore through the mud back and forth, like stitching a wound, to cradle the wreck. Then it could crawl along the scaffolding and attach inflatable pontoons to every inch of the ship.

Woolley could have overseen this process from the ship above. He could have observed scrolling numbers and beeping machines when each phase was underway. And when the operation was complete, he could give the signal for a surface pump to shoot extremely compressed air into the pontoons and stand in anticipation as the seafloor rumbled and the *Titanic* began to rise. The primitive technology existed, and for the right reasons and the right price, it could be deployed, paid for, and put into action. The deep sea was becoming a business, and Woolley's only product seemed more and more to be his childhood devotion to a relic of history. Innovations that might've helped him passed him by completely, a current of ingenuity swirling around him while he sat hopelessly moored in the mud.

° ° °

Getting a person into the deepest parts of the ocean was one of the preeminent challenges of the twentieth century. But once that milestone was reached in 1960 with the team that touched the bottom of the Challenger Deep and demonstrated there was no deeper a human could go, the next frontier was to offer an undersea explorer some personal agency. Rather than simply descend and return to the surface, scientists needed technology to maneuver and explore the peculiar topography of seafloor rocks, mountains, and mud and eventually collect samples of deep-sea animals.

The U.S. Navy had been working on deeper and deeper watercraft to evade detection from the Soviets, who were doing the same.

But both powers focused on warfare more than sharable research, and so the pursuit of science fell largely to nonprofit research institutions on America's coasts. The Massachusetts-based Woods Hole Oceanographic Institution had been established in 1930 at the direction of the National Academy of Sciences to be an epicenter of American ocean exploration. And even though its early days were spent on World War II preparedness and U.S. maritime advantage, by the fifties, the organization renewed its focus on basic underwater science and novel ways to explore it. In 1964, Woods Hole scientists unveiled a manned submarine called the *Alvin* capable of descending six thousand feet—just over a mile. Once proven successful, *Alvin* was immediately conscripted into government service in a high-stakes mission that redefined what it meant to be extraordinarily lucky.

In January 1966, an American B-52 bomber collided with a tanker during a mid-air refueling maneuver above the skies of southeast Spain. As it broke apart, the bomber dropped four thermonuclear hydrogen bombs. Three of the bombs landed on the seaside fishing village of Palomares, a region known for cultivating tomatoes. "I looked up and saw this huge ball of fire, falling through the sky," one of the villagers told a radio reporter. All the bombs detonated their conventional weapons in huge blasts but without producing a nuclear explosion. The fourth fell by parachute into the Mediterranean. Fearing it could explode at any moment, the navy deployed the *Alvin* to comb the seafloor for a nail-biting two months until the bomb was found, retrieved, and disposed of.

Going deeper than a mile, however, proved challenging on account of the human life-support system. Any manned submersible going that deep would need to withstand extreme pressure while also constantly recycling unpressurized oxygen. Meanwhile, the old style of descending, touching bottom, and resurfacing required only a decent cable or air buoyancy tanks, but descending, moving around, and coming back to the exact same spot required global position capabilities before conventional GPS fully existed. The final quandary was propulsion. To reach extreme depths, the *Alvin* and similar crafts needed to be part car and part hovercraft, able to move in every possible direction, but especially up. In the worst-case scenario, a malfunctioning craft unable to arrest its fall would slam into the seafloor.

This exact nightmare happened in 1968. As engineers were lowering the *Alvin* off the side of a navy pontoon off Massachusetts in hopes of studying the tops of seamounts and looking for whales, the support cables holding the *Alvin* snapped and the craft fell in the water. The three-man crew managed to open the hatch and escape, but the *Alvin* filled with water and sank to the bottom. Several minutes later, America's best-equipped search-and-rescue vessel had sunk five thousand feet deep, where it stayed for almost a year until a more advanced vessel called the *Aluminaut* found the *Alvin* and carried it up. It was easier to fix the waterlogged craft than to scrap it, and by 1973, a strengthened titanium hull and a variable ballast system extended the *Alvin*'s reach to thirteen thousand feet, almost exactly the depth of the *Titanic.*

Meanwhile, scientists never got tired of trawling the ocean floor. They took the *Alvin* through the mountain lake waters of the Panama Canal and to the Galápagos Rift, where the discovery of novel marine life in the warm-water vents was thought to explain the origin of marine life and, by extension, all life.

Looking closely, one could see in the steady expansion of deep-sea research the inevitability that the *Titanic* would eventually be found. The technology had caught up with the challenges of the deep sea, and by the summer of 1975, anyone who understood underwater exploration could deduce it was a question of when rather than if the world got another look at history's most famous wreck. Humans weren't ready to search the entire floor of the world's oceans, but coordinates given six decades earlier by a frantic telegraph operator narrowed down the search field to a few dozen square miles. Eventually, one lucky ship captain would discover the wreck, take underwater photos, and return to port with proof.

Woolley was still certain that it would be him, and one day in June 1976, his phone rang and delivered him one last shot at getting it done.

° ° °

Seven decades after the *Titanic* sank, its condition was still mostly a mystery. Broad scientific principles could inform reasonable assumptions about a hunk of steel and wood in a low-oxygen and high-pressure environment. But there remained a major question mark. Except for the conflicting accounts of the survivors, no one knew for sure in what position the ship sank and how it struck the seafloor. Those two factors alone would account for vastly different conditions of the aging wreck.

At worst, the ship fractured into a thousand pieces upon its high-velocity impact with ocean rock. If this was the case, the boilers likely broke loose during the descent and ripped the entire hull apart. The implosion of internal air pockets likely flattened the superstructure, and any three-dimensional components still intact after the deep-sea crash were instantly flattened by the crushing pressure at twelve thousand five hundred feet. After this, the debris itself would be in fine shape, so long as you had a giant rake that could sweep the ocean floor as though picking up the far-flung shards of a broken plate.

An alternate theory was kinder to the ship. Commander John Grattan, a former diving expert for the Royal Navy, held that the ship had sustained only minor damage during its descent and impact. The watertight bulkheads failed while the ship was still floating, which was lucky because it allowed seawater to penetrate every man-made crevice and equalize pressure before exploding in deeper water. Citing the sinking velocity of a missile falling through the water, Grattan believed the ship sank at a docile seven miles per hour, which was diminished further by the "cushioning effect" of compressed water near the seabed. Grattan thought the two-mile-long trajectory that others believed brought the ship to its maximum terminal velocity instead did the opposite. He believed the lengthy free fall gave the ship time to regain its center of gravity to the point that, when it finally reached the seabed, it set itself down gently in an upright position.

Grattan's vision of the wreck was optimistic, but it wasn't outlandish. Imagining the *Titanic* in good condition was crucial to rationalizing the cost, hardware, and time involved in mounting a survey expedition to go see it. This hopeful vision was also helpful to Doug Woolley, who read about Grattan's theory and got in touch to inform Grattan that any survey "had to go through me," as the wreck's supposed legal owner. According to Woolley, Grattan's reaction was deferential. Also according to Woolley, he made no secret of his enthusiasm for Grattan's work so long as he could be associated with it.

The true condition of the wreck was somewhere between the two theories. The lack of knowledge of the *Titanic*'s general condition received a boost in 1976 when a British archaeologist named Keith Muckelroy developed a model for forecasting the breakdown of aging wrecks, including the *Titanic*. Muckelroy's idea, which is still deemed a seminal breakthrough in the field of maritime archaeology, was that a wreck's state is not simply a function of the ship's weight and basic gravity, as Grattan believed. Instead, its condition relied on a lengthy list of factors, including the materials the ship was made of, how specifically it was wrecked, its movement through the water column, the disintegration of its perishable materials, and, perhaps most overlooked, the constant changes of the seabed.

A shorter way of putting Muckelroy's theory was that time was a crucial factor, both in a wreck's breakdown and also toward any attempts at excavation. Muckelroy published this finding after discovering the *Kennemerland*, a merchant ship of the Dutch East India Company that wrecked off the Scottish isles in December 1664. After nearby islanders helped themselves to the valuables floating in wood cases, the *Kennemerland* sat submerged for three hundred years. Muckelroy found the wreckage in 1971 and learned, at age twenty, that good news shouldn't always be shouted from rooftops, and that announcing the *Kennemerland* discovery would effectively be its death sentence. "Only complete excavation can preserve the site from looting," he wrote at the time, sober about the fact that unlike land-based archaeological sites, which could be easily guarded by people, barriers, and guns, protecting underwater sites from treasure hunters and memorabilia seekers was almost impossible.

As a matter of protection, the *Titanic* had the best security a shipwreck could dream of. The most expert hunters couldn't find it. Advanced divers couldn't pilfer it. Everything that sank with the *Titanic* was likely still with it.

But something about the current was changing. With the arrival of the 1980s, a clock somewhere had started to wind down. And with it, the reality that the wreck's time alone in the quiet dark of the deep sea was coming to an end.

Chapter 9

# PEOPLE THINK SINKING SHIPS IS EASY

In late 1978, John Grattan, the senior Royal Navy diving expert, was the odds-on favorite for the man who would find the *Titanic.* His detailed theories about the ship's status on the seafloor conveyed an advanced understanding of the complexities of the wreck. Grattan believed the *Titanic* was exactly twenty-two miles from the location given as its final coordinates. The precision of his declaration suggested immense confidence, and without anyone able to present compelling counterevidence, Grattan's speculation was as good as fact.

Grattan also had a track record of success. Several years prior, he commanded the HMS *Reclaim*, a British salvage ship that combed the Irish Sea to recover the wreckage of a crashed Aer Lingus plane. Not long after, he recovered a downed helicopter in the northern waters off Norway. Both operations used sonar crafts and underwater cameras, which earned him technical credentials.

He also had experience searching for one of the most elusive and horrifying shipwrecks of all time. The cargo vessel *General Grant* had been sailing from Melbourne to London in 1866 when it hit cliffs in the notoriously perilous waters south of New Zealand. Most of the eighty-three passengers died, but the story of the *General Grant* grew darker in its aftermath. Thirteen passengers fought fierce waters and made it to a remote island, where they lived as castaways and survived on potatoes and seals for a year and a half in some of the most punishing weather on earth. Eighteen months later, only ten remained when a sealing brig passing by saw their frantic signals and took them to New Zealand.

The *General Grant* was worth salvaging not because of the toll of human lives but because it was rumored to be holding at least twenty-seven hundred ounces of gold, worth, in today's currency, more than $5 million. Search and salvage attempts began just two months after the survivors were rescued in 1868 and continued until 1916, when scavengers stopped trying. After more than a dozen ships had visited the site and divers swam more than forty nautical miles around the area and returned empty-handed, treasure hunters mostly gave up.

Half a century later, Grattan gave it another shot, and in 1975, he actually found it—or at least he found *something*. He returned to England with pieces of mid-nineteenth-century wood, bent metal, and what seemed like shards of glass that he believed to be from the *General Grant*. Notably, he found no gold. Later surveys revealed that Grattan's claim of the *General Grant* was actually a French ship called the *Anjou*, which wrecked in 1905, likely while looking for the *General Grant* as well. To this day, the wreck and its gold are still waiting to be found.

Several years later, Grattan's résumé and his interest in the *Titanic* caught the attention of a millionaire pharmaceutical tycoon named James Goldsmith, who, upon meeting Grattan, agreed to finance an underwater survey at the exact site Grattan believed he would find the *Titanic*. Goldsmith's interest was partly for the sake of scientific inquiry, but mostly as an investment in media and marketing. Before he inked the deal with Grattan for the survey, he brought in Woolley too, in case Woolley's claim of salvage rights actually held up in court.

Goldsmith was relatively new to wreck hunting, but he was a familiar character in the history of high-profile scientific quests: a rich man at the top of his industry who had grown bored. Money had swelled his ego, which had fueled his belief that the person who found the *Titanic* would be an international celebrity. Goldsmith figured that the media's reaction to his plan to survey the *Titanic* would be equal to a private company trying to put a man on the moon. He'd put up the money so long as someone else was willing to do the work.

As it happened, Goldsmith had approached Woolley at the perfect time. By that point, Woolley had shed everyone on his previous team, including the boy mascot and the young man he found lying under the car. The group was riddled with mismanagement and sniping, much via written correspondence that had the effect of enlarging small tiffs into lengthy distractions. Woolley was also growing impatient. He had been promising immediate action on his plan for more than a decade. "I got the sense I needed to increase the visibility of the operation, and to do that I brought on more serious people," Woolley recalled years later.

Goldsmith offered Woolley no money—he likely realized he didn't need to—but he did promise Woolley that his name would appear on all marketing materials for the survey and he would be included in the highly publicized celebrations of its success. Faced with the existential dread of receiving nothing at all, Woolley saw this as a no-brainer.

The fairness of the arrangement was a matter of perspective. Woolley saw himself as the nucleus of the operation and was excited to partner with a well-known millionaire. After ten years of trying to drum up the money, he had finally found the limitless underwriting he needed.

Goldsmith, however, saw Woolley as an insurance policy. Success required people who knew what they were doing, and in addition to Commander Grattan, Goldsmith brought on an accountant named Clive Ramsey to manage the finances and an entrepreneur named Philip Slade, who could coordinate the mechanics. Woolley was welcome to do interviews about the operation and hype its inevitable success. But he was pointedly not invited aboard the boat that in the spring of 1980 would go to the wreck site and tow a series of sonar rigs more than two miles underwater until a series of beeps indicated the wreck was found.

Before embarking on the survey, Grattan locked himself in his house and double-checked every possible detail known about the ship. He read dozens of books about the disaster and also the 1912 incident report by the British Board of Trade, which was filled with painstaking detail about everything from the design of the ship to its construction, its equipment, the precise ways the lifeboats were filled and launched, the contents of all wireless communication, official orders, passenger biographies, crew credentials, instances they were trained for, and matters of on-board discipline. After that, he turned to marine science. He sought out experts on ocean pressure, temperature, marine life, and photodynamics. When people said that the deep sea was dark, did they mean dark like walking to the bathroom at night or dark like being blind? He examined tides and ocean currents and learned about celestial navigation and snowfall in Greenland. No morsel was too minute for Grattan, who thought that if he knew enough, he could almost *become* the *Titanic* and picture himself in its place.

Grattan mapped out his plan in rough drawings and chicken scratches. A sonar-powered submersible would be towed behind a boat and scan the seabed from both sides. The sonar vehicle would relay pictures back to the search ship, where the team would look for signs of debris. If the crew identified a book or candlestick or windowpane, the rig would perform another pass, and upon confirmation, it would shrink the search area until the hull came into view. Once the sonar rig found the wreck and sent down a transponder to mark the spot, the crew would release a remote-operated vehicle equipped with lights to illuminate the wreck and take pictures. Then the remote vehicle would open like a Russian doll to release a smaller craft. The larger portion, called "the mother," would return to the surface with the early photographs. The smaller craft, "the baby," would enter the *Titanic* to film its interior.

Grattan believed the images would be earth-shaking. The British writer Joseph Conrad had once referred to the *Titanic* as the "marine Ritz," after the tony London hotel. Seventy years later, Grattan imagined the ship's fine finishes and ornate flourishes had spared it from extreme ocean degradation. But beyond that, all Grattan could do was hope for the best.

° ° °

Optimism in the face of uncertainty is always better than the opposite. But the condition of the ship and any of its contents rested on the unknown condition of the seafloor.

Expeditions for decades generally confirmed Darwin's desert theory of the oceans—that underwater environments were the same everywhere, and, thus, identical. Scientists pulled up samples of seafloor, and the sea cucumbers and mollusks they collected in the Pacific appeared to be the same as ones in the Atlantic.

This began to shift in the late sixties, when two scientists from Woods Hole replaced their large-holed nets with fine mesh that could grab more sediment. On their first drag in the North Atlantic, they pulled up three hundred sixty-five unique species.

In 1975, one of the marine scientists, Dr. Robert Hessler, made another seismic advance in deep-sea biology. In his lab at Scripps Institution of Oceanography in San Diego, he built a box with a movable bottom that, when lowered off the side of a boat, could scoop up as much as three cubic feet of ocean mud. His first test with it revealed mind-blowing levels of biodiversity and density. Amphipods found at one site were visibly different from ones a few miles away, to say nothing of the presumed colonies of microorganisms at each site. "Most species are rare, being encountered only once," Hessler reported. He took the device to the nearly five-mile-deep Philippine Trench and to the seafloor under Antarctic ice and pulled up never-before-seen one-celled organisms that pulled nutrients from the water *and* dug roots into the mud—like a hybrid of a human and a tree. Several years later, he made an even more astonishing discovery. While anchored near the Galápagos, Hessler realized that living two thousand feet deep were dense numbers of novel worms, mussels, fish, crabs, clams, anemones, and shrimp clustered around a hole where hot water was coming out, some of it boiling. The water was heated by thermal vents in the seafloor and was high in hydrogen sulfide—the unmistakable smell of a rotten egg—which fed bacteria that in turn nourished the entire food chain. Life wasn't only stewing in the least likely place on earth; it rivaled the rich biodiversity of tropical rainforests.

What effect would this wildlife have on deep-sea wrecks? Grattan didn't hear of Hessler's findings, and if he had, he would likely have dismissed them as irrelevant. Tiny invertebrates living around a wreck had no bearing on the wreck itself. Crustaceans couldn't dine in the *Titanic*'s underwater salons or sleep in the officers' cabins. Grattan, Goldsmith, Woolley, and every other person who had staked their reputation on finding the *Titanic* clung to the belief that the ship had been preserved in nearly the same condition as the night it sank. To admit otherwise would be to acknowledge that the object they had in their sights might not be an object at all, but a crumbling mess of lifeless debris. And once you pulled that string, the rest would begin to unravel.

∘ ∘ ∘

The environmental awakening of the 1970s, brought about by Rachel Carson's *Silent Spring* and her book before that, *The Sea Around Us*, challenged a rethinking of the oceans not as an infinite ground for fishing and dumping but as a biological system supporting all life on earth. Carson never used the word *pollution* in her second tome—she didn't need to. By illustrating the wonders of the currents and climate, the way the ice caps power a conveyor belt that regulates weather, nutrients, and migratory patterns, the subtext was clear that messing with it came at humans' collective peril.

The awakening also brought a shift about wrecks. For each study performed on how ocean life was damaging wrecks, there were fewer that flipped the question to ponder whether the ever-increasing number of wrecks were in fact damaging the ocean.

Protecting the ocean purely for the ocean's benefit has never been popular among humans. The closest we've come has been to ask whether supporting the ocean could somehow benefit us. By 1970, the answer was yes, and conveniently, it required *more* wrecks.

Japanese fishermen knew for centuries that man-made reefs could help them grow kelp and establish new colonies of fish. Americans warmed to the idea in 1916, first in the waters off New York, where a local fishing union filled butter tubs with cement to grow coral that would attract fish. Half a century later, artificial reefs had grown into a powerful economic force, not only for fishermen but also for shipbuilders who had to contend with the inconvenient matter of what to do with an aging ship. The Alabama Department of Conservation initiated the first official reef-building effort in the Gulf of Mexico in 1954 by sinking four aging ships that would otherwise have been dismantled and thrown in landfills. It was far easier to flush their tanks and remove all petroleum, oil, and lubricants and send them out of sight and mind. By 1958, California went a step further and began to sink old ships, old cars, and anything at all to lure fish—but really to lure fishermen.

Conveniently, as early-century ships were reaching their operational limits, there was suddenly an abundance of ships qualified to become wrecks. Florida became the epicenter of reef-building. It had a lot of boats, a lot of shoreline, and warm water hospitable to coral. By 1970, Florida officials were sinking a new vessel almost every week. And when there weren't enough old boats for this surprisingly easy and satisfying form of environmental protection, they granted permits to sink old toilets, water heaters, and hollowed-out cars. If all that coral needed was a surface to grow on, then it stood to reason that metal, porcelain, or rubber would be better left in the ocean than in a landfill.

Things finally got out of hand in 1972, when giddy Florida officials dumped two million old tires in coastal waters as part of a plan to build a field of coral to rival the Great Barrier Reef. The quantity was so large it not only smothered existing marine life, but the tires weren't heavy enough and roamed with the current like tumbleweeds across the seafloor, killing existing coral and demolishing the ecosystems they supported. As recently as 2015, divers were still pulling thousands of tires out of the Gulf of Mexico each year.

∘ ∘ ∘

Sinking is the very worst thing that can happen to a ship. But eventually, there arrives a day when sinking isn't so bad and can even become so helpful that people will pay to do it.

One day in the fall of 2020, I caught up with Tim Mullane, who runs a ship-sinking company in Virginia called Coleen Marine, named after his wife. The couple, along with a small crew, operate a funeral parlor for old ships, preparing them for the underwater afterlife.

"People think sinking ships is easy," Mullane said in a raspy voice while driving from one job to another. "Like you just leave a window open and eventually the thing goes down."

That happens, and ships have certainly sunk for more boneheaded reasons, but scuttling a vessel to make an artificial reef is a maze of endless regulation, physical labor, and ocean dynamics.

If done correctly, an artificial structure can mimic the topography of a natural reef. In warm water especially, a wreck acts as a place for coral polyps and floating planktonic larvae to settle. Once they do, they begin to attract crustaceans and larger invertebrates that in turn attract fish. Fish can theoretically leave a reef whenever they want, but most don't, a lesson illustrated vividly in *Finding Nemo*. This makes artificial reefs placed near natural reefs especially helpful to expand habitat and enlarge populations.

That's if it's done right. After the reef-making bonanza of the seventies, the pace slowed considerably after new research suggested creating healthy reefs wasn't as simple as just dumping an old boat. Vessels are composed of hundreds of materials and thousands of parts that can end up in the digestive tracts of sharks or turtles. Attracting too many fish to a spot they wouldn't otherwise go can make them an instant target to predators, including humans. And as with real estate, location is everything.

First, Mullane told me, you've got to clean the vessel. That means pulling out the electrical motors, wiring, lead, insulation, gaskets, carpet, and wood, as well as stripping off the paint, which can disintegrate and find its way up the food chain. Draining the fuel is obvious, but you have to get all of it. Even a half pint of diesel will make an eight-square-mile sheen on the water. Once the vessel is reduced to a carcass of steel, the ship is prepared for wreck life, which requires cutting big holes for fish or divers to swim through and ensuring smooth water flow so a strong current doesn't pick it up and roll it like a motorhome in a hurricane. Mullane's team simulated "swimming" through it on land, smoothing down hard edges that could snag a diver's equipment.

When it comes to sinking, the ocean does most of the work, but amid so many variables like the wind, ocean current, and sinking velocity, it's important to do it fast. You can drill holes beforehand and plug them until sinking time—a square foot apiece is usually fine—but the more the better to ensure the vessel goes down exactly the way you want, either bow first or stern first or down on its side. You can do everything right and a ship will still invert itself at the last moment. Oddly, this happens frequently in Delaware due to underwater currents. Sometimes you can use ropes to jostle it around after the fact. Sometimes you can't.

I fanboyed a little and told Mullane it was cool to watch his ship-sinking videos online. He was quiet for a moment and then said he rarely got compliments. In seventeen years he'd sunk more than a hundred ships, most of them 180-footers or bigger, almost all the general public will never see again. He sounded tired.

"I used to find it exciting to watch a ship sink," he told me. "Now I'm just relieved the job is over."

Once a ship is gone, it belongs to the state government or environmental group that hired him to sink it. Which is another way of saying, it's not his problem anymore.

° ° °

Anyone watching the feverish launch of ships in the early-twentieth-century era could have anticipated that six or seven decades later, the same ships would become limping vessels covered in rust and barnacles, if they were still floating at all. It's unlikely the *Titanic* would have made it this far; its sister ship the *Olympic*, which was launched in 1910, reached the end of its efficient operation in 1937, when it was sold to a man whose sole interest was to provide ship-breaking jobs for men in need of work.

Some ships outrun forced retirement but end up like fading musicians booked for smaller and smaller venues. The *Queen Mary* was launched in 1936 and spent an illustrious half century carrying rich vacationers and World War II soldiers. In 1987, it was bought by the City of Long Beach, California, where it was stripped of all seafaring capability and turned into a floating hotel. Since then it has changed hands a dozen times. Almost every operator declared bankruptcy after realizing their grand visions for restoring the aging liner were dwarfed by the costs.

The same is true for the 1950s-era SS *United States*, once the largest and fastest American ship. For the past twenty-five years, it's been falling apart on the Delaware River in Philadelphia as a literal lounge act. Having once carried Harry Truman, Marilyn Monroe, and Marlon Brando, the *United States* is now best known as the backdrop for the nearby Ikea cafeteria.

I grew up in Southern California and still remember the first time I saw the *Queen Mary*. My family was driving nearby, and my mom pointed it out as if she had spotted the Hope Diamond or the Taj Mahal—"*Wooow, there it is.*" It's obvious now that the *Queen Mary* is moored under its own nostalgic weight, like a photo album you can't throw away because it once belonged to your grandmother. It once meant something, and retiring it feels like a violation of the past. But keeping the *Queen Mary* on life support only delays the inevitable fate of all ships to be either sunk or scrapped.

Shipbreaking is an industry almost as big as shipbuilding, but with a major geographic difference. Ships are made to withstand extreme pressures, so they're built with strong and often toxic materials like lead and asbestos. Owing to costly safety regulations in developed countries that eat into already slim profit margins of deconstructing old ships, most breaking occurs in places like Bangladesh, India, and China, where workers have little, if any, protection. In 2014, *National Geographic* called shipbreaking "one of the world's most dangerous jobs."

Ships are broken down in the reverse order of how they're built. First workers remove all fuel, oil, and other fluids. Any valuable equipment like the engines, navigation computers, and electric generators are pulled out and resold. Then hundreds of workers with acetylene torches swarm the skeleton to pull off large sheets of steel, which porters carry to furnaces, where they're melted down into alloys like rebar to become the bones of new buildings or sometimes new boats.

A reporter friend of mine went to visit a shipbreaking yard in Bangladesh several years ago and returned with stories that still haunt me. The entrance to one of the yards was so guarded from outsiders that my friend had to hire a fishing boat to take him over water to the disemboweled ships. Sparks rained down on him, and he narrowly avoided being struck by a piece of falling steel. He met a family with four sons who were all victims of shipbreaking trauma. The first son worked as a helper for two weeks before he saw a man burn to death when his torch caused an explosion. The second brother was cutting off a large section that fell on him and killed him. The third brother slipped through a hatch on a tanker and fell ninety feet into the ship's hold; he survived because water had pooled in the lower decks and broke his fall. The last brother, with few choices for work, still worked on the ships. Most of the men my friend met had deep scars. Several were missing fingers or were blind.

Breaking down smaller boats is easier and safer, but as is common in recycling, there are parts that can't be resold or reused. They're just trash. California once had such a problem with people sinking their own old boats and causing small oil spills and navigational hazards that officials started imposing a $3,000 fine for self-disposing plus the cost of removal. Recognizing that people still do dumb things, particularly in California and especially in the dark of night, the state started a no-questions-asked boat disposal program, like dropping a baby at a police station. This helped, but one boat disposal expert I spoke with said people still sink their own boats anyway, usually for insurance money, and often in the deepest water they can manage.

Most get away with it. Boats with gold are hard enough to find. Who's going to go look for a worthless old pontoon?

∘ ∘ ∘

The ledger for John Grattan's proposed survey worried the millionaire James Goldsmith. The operation included fixed costs like renting a boat, borrowing sonar equipment, and several weeks of salary for the crew. But then there were variable costs—the insurance for the boat, limitless fuel charges, faulty equipment replacements, and weeks or months in overtime pay if the wreck was slow to turn up. And odds were, it would be. Goldsmith could absorb any overruns, but no one becomes a millionaire by making bad bets, and making a high-stakes guess in a swirl of uncertainty had all the signs of a bad bet. Goldsmith offered to pony up a third of the money if Grattan could raise the other two-thirds in sponsorship and media deals.

This brought up the question of what sort of returns investors could expect—if any at all. Most salvage companies can mobilize the crew and equipment to recover a buried treasure, but the *Titanic*'s worth was almost entirely in monetizing its media splash. There was no doubt that the reemergence of the *Titanic* would be international news, but there was no guarantee that the person who found it would be the one to cash in. Demonstrating this lopsided financial equation, two companies with almost unlimited coffers, Disney and National Geographic, joined forces for a *Titanic*-searching expedition in 1978 in hopes of turning the ship's rediscovery into a cultural event with limitless tentacles: movies, books, museum exhibits, amusement park rides, consumer products, and looping TV specials. The companies hired a salvage firm and a team of deep-sea engineers. But when the price tag continued to rise on account of multiple ship rentals, fuel, interminable room-and-board for the crew, advanced deep-sea equipment, high-paid technicians to operate it, and insurance to the moon and back, *all with no promise they'd even find anything*, the bosses scrapped the idea and walked away.

During Grattan's public appeal for sponsorship from a media company, Doug Woolley was pushed to the background, where he was willing to sit, albeit uncomfortably, in exchange for a chance at Grattan's success. But as Grattan's underwriting was failing to materialize, Woolley felt the familiar disappointment of another attempt ending before it began. Goldsmith called Grattan, who relayed the news to Woolley that the project was done. In the clearest sign of defeat, Goldsmith closed down the glossy magazine he'd bought, *Now!*, through which he intended to break the news of the discovery of the *Titanic*.

This left Woolley back at square one, where most men who have weathered as many defeats might have given up entirely. But Grattan's failure was almost a relief for Woolley, because once he got past the sting of another faceplant, he realized he was back in his comfort zone: on his own, with nothing to do but regroup, rethink, and give it another go.

° ° °

I got lost one afternoon down a rabbit hole of shipping collisions and accidental sinkings. I wanted to learn about a specific ship, the tall ship *Bounty*, which sank off the coast of North Carolina in 2012 due to—and this is according to the National Transportation Safety Board—"the captain's reckless decision to sail the vessel into the well-forecasted path of Hurricane Sandy."

After watching news coverage about the incident, the number of "suggested videos" that followed made clear that this sort of thing happens surprisingly often. There were videos of enormous ships colliding in slow motion and small ships with inexperienced captains who dock the way a fifteen-year-old parallel parks. I watched a twenty-two-minute video titled "Biggest Container Ship Accidents in the 21st Century" and another called "Cruise Ship Crash Compilation #2." At one point my wife came home, and I called for her to come watch a cruise ship ram into a dock. We watched with our dog—six eyes glued to the screen—as the MSC *Opera*, a sixty-five-thousand-ton ship capable of carrying twenty-five hundred people, plowed into both its concrete dock *and* a smaller riverboat in Venice in the summer of 2019. Five people were injured, which seemed like a small miracle. Think how frightening it must have been to be standing on the dock when an out-of-control, skyscraper-size piece of steel was coming at you.

Once you start looking, the internet is full of what's inventively known as shipwreck porn. (There is also the opposite: online communities that come together around their *fear* of wrecks, a condition known as *submechanophobia*.) Photographers have filled entire careers taking photos of beached boats under moonlight or of regatta races gone wrong or of ships that sank in areas where there isn't water anymore, giving aging steel structures a dusty post-Soviet facade. And families fill their entire Sundays going to look at local wrecks like the USS *Inaugural* in the Mississippi River that go from above water to below every few weeks with the rise and fall of the river. One of the strangest wreck varieties I found started with the HMS *Victoria*, a British battleship that sank in 1893 and was discovered near Tripoli in 2004 to be perfectly vertical, like a pencil standing on the seafloor. This is rare but not unheard-of. If a ship's cargo is front-loaded or its propeller continues spinning as it sinks, a vessel can ram itself fifty or more feet into mud.

Shipwreck forums tend to be active every hour of every day with enthusiasts discussing their favorite wrecks. You might think sunken ships are fairly straightforward. But the debates about wrecks are almost as abundant as wrecks themselves. What grade steel has the best buoyancy? (Consensus is grade 5.) Are captains cowards who flee sinking ships before passengers? (Debatable, but yes.) Can a boat actually be unsinkable? (Yes, a Florida-based company uses a chainsaw in advertising to demonstrate that its foam-core hulls can stay afloat even after being cut in half.) My favorite was a lengthy argument over where on earth is the worst place to be wrecked. There is a definitive answer: Point Nemo in the South Pacific, the farthest spot on the planet from a major landmass. Add in rough seas, near-polar temperatures, and the fact that Point Nemo is the unofficial dumping ground for satellites falling out of the sky, and even the most devoted wreckheads will agree, don't go there.

What all discussion boards, fan pages, and shipwreck email Listservs have in common is that all of them eventually return to the north star. You might be discussing shipping collisions, of which there have been millions over millennia, and someone will bring up the iceberg. You can find engineers debating how to lift up old ships, of which there are hundreds of thousands, and someone will always ponder why we never resurfaced the *Titanic.* How are ships built? "*Well, look at the* Titanic . . ." Why do ships sink? "*Well, in the case of the* Titanic . . ." How dangerous is shipwreck diving? "*Good luck diving down to the* Titanic." I posted in an online forum one day for people to nominate their favorite shipwrecks. Eight of the first ten people simply posted "*Titanic*," as if it required no other explanation. As a character, it reminded me of Genghis Khan or Hitler, a black hole of history through which all events intersect.

A more apt comparison, however, might be to Kim Kardashian or Donald Trump, famous for being famous and savvy at leveraging that fame at opportune moments. Several times in the twentieth century when the *Titanic* was at risk of fading from public view, new media productions introduced it to new generations, which fueled new rounds in the zeitgeist. You might think that can happen only so many times, and that now, after more than a century and hundreds of movies, expeditions, court cases, and books—including this one—that the *Titanic* is on the wane. But according to Google, the ship remains at "peak popularity" as one of the world's most popular search terms almost every day.

Understanding how deep the *Titanic* has wedged itself in the world's collective consciousness might help explain how far the ship had burrowed deep into Doug Woolley's soul.

"Time is working against us if we want to get the job done," Woolley began writing to friends in 1979. "We must work harder and faster and I'm doing all I can to hold it all together."

As it happened, the most important partner Woolley could find actually found him first. His phone rang one evening and on the other end was a Texas billionaire with a thick southern drawl.

"I'm looking for a man called Woolley," the voice said. "You know that son of a bitch?"

Chapter 10

# A HEIFER CORRALLED IN A BOX CANYON

Well, you won't get very far talking to me like that," Woolley told the man calling.

"Now, listen," the voice said. "I'm looking for the *Titanic* and I want your help."

There was a silence. Woolley smiled.

"How can I be of service?"

Jack Grimm had no time to waste, certainly not in making small talk. Brash and profane, if Grimm had an idea in mind, he drilled to it quickly and used equal parts money and bluster to cut to the chase. That was how he had built his multimillion-dollar fortune and what fueled him on his quixotic hunts for the Sasquatch in the Pacific Northwest, the Abominable Snowman in Nepal, and the Loch Ness Monster in Scotland. He even financed an expedition to the north pole at one point to "prove" the fringe geologic theory that the earth was hollow.

Grimm was a prospector of all things. He had begun at age eleven, when he heard a rumor that there was treasure buried in a nearby riverbed. Instead of digging out the river, he bought a few sticks of dynamite from the hardware store and turned up a few arrowheads and an old frying pan. As a young man, he was enchanted by a friend's father, who had been a rich oil prospector and told the boys stories of high-on-the-hog riches based on little more than a good guess. After fighting in World War II in Okinawa, Grimm went to study geology at the University of Oklahoma. When job offers came, he turned them down to work on his own. The first well he drilled in Oklahoma struck oil and nearly made him an overnight millionaire. He used the money to buy land he thought might have gas under it. The thrill of betting big and winning often enough was intoxicating to Grimm, so much that when he married his wife, Jackie, in 1947, they spent their honeymoon in California panning for gold.

Success was often a matter of interpretation—particularly Grimm's. In 1970, he went to Turkey to hunt for lost remnants of Noah's Ark. No serious scientist thought he actually found it when he returned with a piece of carved oak pulled from the side of Mount Ararat. He suspected the wood was four million years old, and that alone was enough to prove it was Noah's. "This is the ark, that's my story, and I'm going to stick to it," he said.

Eventually, the quest for higher and higher stakes put him on a search for the most famous shipwreck still waiting to be found.

Grimm was just as eccentric as Woolley, possibly more, but he was better funded, and that made all the difference. A man of means, he had little trouble corralling a team, which he began in the summer of 1979, and chartering a boat, which cost $10,000 per day. Trying to emulate Grattan from several years before, Grimm did a cursory study of weather patterns and ocean currents to know what to expect. He consulted survivor accounts and time logs to draw a one-hundred-thirty-square-mile search area and sought the advice of marine experts about the challenges of seeing underwater. Unlike Grattan, Woolley, or anyone else who had ever tried or wanted to try, Grimm was rich enough to personally invest in the best equipment possible, which included sonar and side-scanning cameras to trawl the seafloor.

His organization was flawless, but by the spring of 1980, several weeks before his summertime expedition was to depart, he began to wonder if there could be legal trouble if he actually found the ship. A lawyer looked into it and reported back to Grimm that the man who apparently owned the *Titanic* was living in a small apartment on the east side of London.

Doug Woolley liked Grimm immediately. The fast-talking Texan was an American stereotype. Once they got talking, Woolley realized this might've been the partner he'd been looking for for years. Grimm had a background in geology, a sober understanding of what was required to find the ship, and a nearly limitless budget to do it. Most striking, though, was Grimm's foul language.

"I've never heard a mouth on a man like that in my life," Woolley later said.

Grimm called Woolley under the guise of partnership, but his real goal was to charm Woolley sufficiently to avoid a messy media squabble once the ship turned up. Discovering the *Titanic* would trigger a tsunami of headlines, which could crash quickly if the story shifted from Grimm finding the wreck to Grimm getting sued for disturbing a hallowed wreck site that ostensibly belonged to someone else.

Woolley was hardheaded, ego-fueled, and singularly focused on *him* being the one to do it. But he was also easy to flatter. Woolley had grown accustomed to the haters, the doubters, and the endless mockery, including from his own family, who had all but excommunicated him. He hadn't talked to his sister, Anne, in months, which turned into years. Making friends came easy, but the people drawn to Woolley were often hangers-on, misfits, or lost boys of their own. What he needed were people as smart and savvy as him—or even smarter, so long as they didn't overshadow him. This longing created the opening for a man of money and status to cajole him with compliments.

Grimm might have sensed this, or he might've seen in Woolley a kindred spirit, a man endlessly misunderstood but with the drive and desire to prove everyone wrong. Grimm stayed on the phone with Woolley for more than an hour. They discussed Woolley's past plans, his near misses, his band of ragtag associates. Grimm told him profanity-laden stories about his trip to Turkey and his home in Abilene, Texas, and the time he nearly went bankrupt drilling twenty-four holes until he came to the last one, the twenty-fifth, and it turned out to be a gusher. Woolley loved the way Grimm dropped Texas-isms. Wildcatting for oil—the slang for prospecting—was no different than finding a ship, Grimm said. He was riding a gravy train with biscuit wheels. When they succeeded, that'd throw their hats over the windmill. Woolley laughed and laughed, and when Grimm asked if he had Woolley's blessing, Woolley told him to go find the damned ship.

Grimm wasn't all hat and no cattle, as a true Texan might say. He had gone further than Grattan or any other hunters in at least one way, by solving the *Titanic*'s long-disputed CQD mystery. The ship gave its well-known emergency coordinates as it was sinking, but a *Titanic* buff in the 1950s had noticed a bizarre quirk about the ship as it sank. Unlike an airplane that moves quickly through time zones, a ship crossing from England to America had to account for losing five hours over the course of six days. It was crucial this be done precisely, since time was an essential variable when determining speed and, by extension, the ship's position. The best westbound ships could do at the turn of the twentieth century was to manually move the clock back twenty minutes every four hours. The process proceeded smoothly for the *Titanic*'s first days at sea, but at midnight on April 15, during the pandemonium of sinking, had anyone bothered to change the time? If not, when the navigator went to pinpoint the precise coordinates to summon help, he may have miscalculated the position by nearly ten miles.

Looked at one way, ten miles is a small margin of error on a journey of more than three thousand. Like most ships since the exploration age of Columbus and Magellan, the *Titanic* navigated using dead reckoning, which required ships to laboriously track their position in relation to a fixed object, like a port or, less precise, the moon. The classic exam question about a car driving from Nashville to Phoenix going sixty miles per hour is a matter of measuring distance based on speed. But the equation grows more complex on the open ocean when accounting for wind gusts and water streams, and even more complicated when a ship has lost its propulsion and bobs in a nonlinear current for more than two hours while sinking.

For Grimm, the ten-mile difference implied a plausible search area of more than a hundred square miles, or roughly the size of Sacramento. Coupled with the knowledge that the ship was more than two miles underwater, the searchable area expanded to two hundred cubic miles, or more than two hundred quadrillion gallons of seawater. Grimm wasn't looking for the ship in the middle of the water column—everyone knew it was sitting on the seafloor—but submerging a camera and sonar rig more than two miles deep meant it had to be towed behind the search vessel on a cable nearly four miles long. Using the Pythagorean theorem of right triangles, this implied the search ship would need to travel at a constant speed of two knots, or barely over two miles per hour, to keep the underwater rig exactly forty feet above the seafloor. Speed up and the rig would rise too high; slow down and it might hit bottom. Get it just right and you still had the major geometric hurdle that once the lure detected a sonar or magnetic disturbance, the area would be two miles behind the ship, which would require at least six hours and an extremely wide U-turn to pass over the exact same spot again.

This was tedious in 1980, and it still is. Any ship sitting in the deep ocean cannot be seen by satellites, metal detectors, or any surface-based technology, leaving sound waves from a sonar rig as the most reliable way to see what's below. And that's if you're lucky. In 2013, a salvage company based in Mauritius called Deep Ocean Search broke the depth record when it found the British wartime liner SS *City of Cairo* more than seventeen thousand feet deep—almost a mile deeper than the *Titanic*—in the South Atlantic. The *City of Cairo* was sunk by German torpedoes in 1942 and became one of the most awful maritime tragedies of World War II, not because the passengers went down with the ship, but because they floated in lifeboats for three weeks before rescuers arrived, finding ninety of them dead. Two hundred passengers and crew survived, some floating in lifeboats for over a month. Adding to their trauma, three of the survivors were picked up by a German blockade runner ship called the *Rhakotis* that was on its way from Japan to France. When the *Rhakotis* was off the coast of Spain, a British cruiser fired a torpedo and sank it, leaving the *City of Cairo* survivors on board fleeing their second wreck in six weeks.

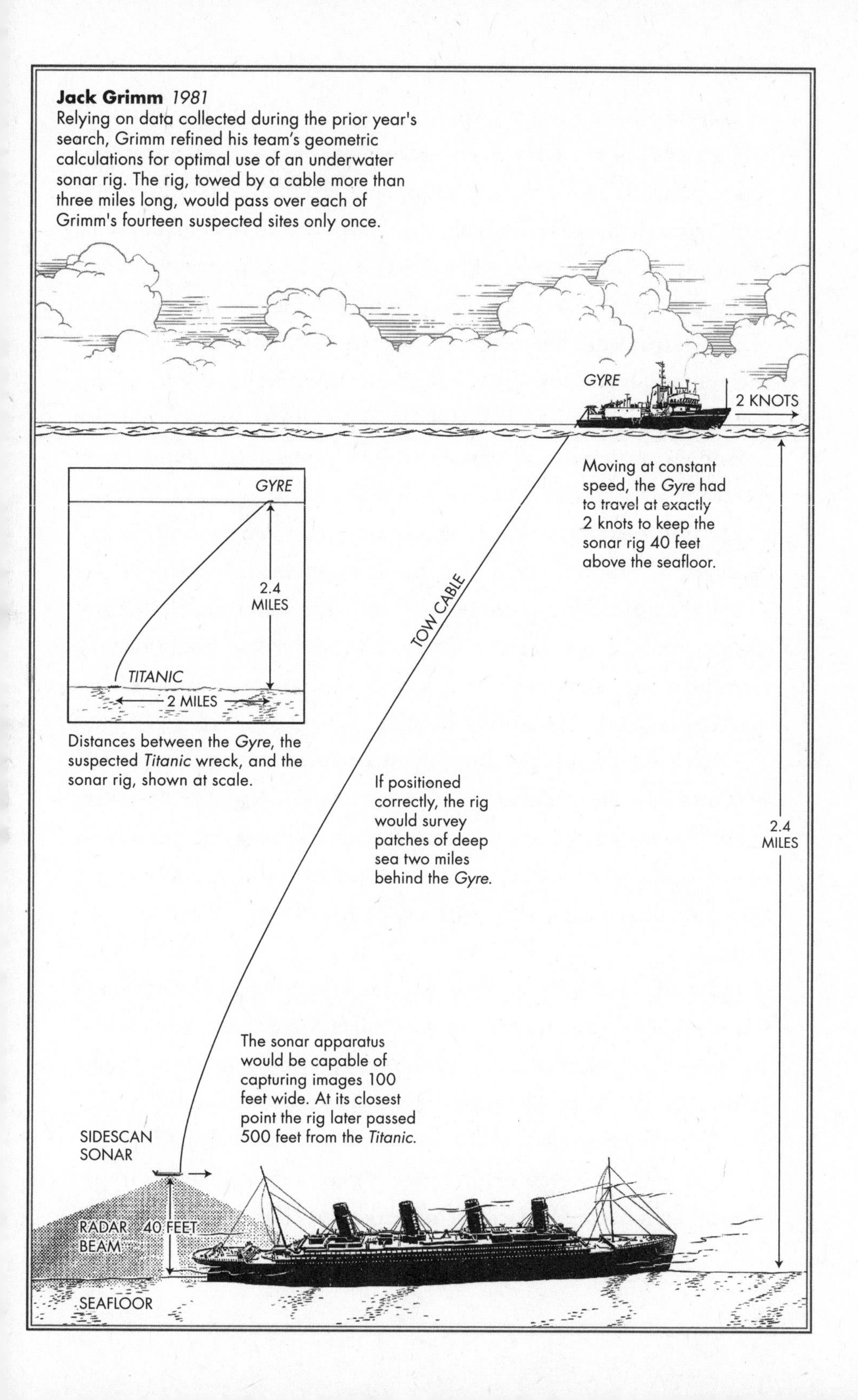
**Jack Grimm** *1981*
Relying on data collected during the prior year's search, Grimm refined his team's geometric calculations for optimal use of an underwater sonar rig. The rig, towed by a cable more than three miles long, would pass over each of Grimm's fourteen suspected sites only once.
GYRE
2 KNOTS
Moving at constant speed, the Gyre had to travel at exactly 2 knots to keep the sonar rig 40 feet above the seafloor.
GYRE
2.4 MILES
TITANIC
2 MILES
Distances between the Gyre, the suspected Titanic wreck, and the sonar rig, shown at scale.
TOW CABLE
If positioned correctly, the rig would survey patches of deep sea two miles behind the Gyre.
2.4 MILES
The sonar apparatus would be capable of capturing images 100 feet wide. At its closest point the rig later passed 500 feet from the Titanic.
SIDESCAN SONAR
RADAR BEAM
40 FEET
SEAFLOOR

Seventy years later, searching for the *City of Cairo* was worthwhile because it went down with a hundred tons of silver. For weeks, searchers made enormous elliptical turns around the area, towing a cable more than five miles long, then pulling it up on a giant winch and analyzing the images. "When you see something that could be a shipwreck, nine times out of ten it's not," John Kingsford, a member of the Deep Ocean Search team, told me. "It's almost always just a rock outcrop that looks like a ship." His team eventually found the lost *Cairo* a few hundred miles south of the island of St. Helena and sent down an unmanned vehicle to recover some of the bounty, leaving the team with a haul of $50 million.

In July 1980, Grimm was attempting the same method. Fueled by nearly $1 million—most of it from his own pocket—Grimm corralled a team of twenty-three scientists and engineers, three filmmakers, and a book author. They set sail from Port Everglades in Florida to find and film the wreck of the *Titanic*. Several weeks earlier, Grimm had learned of a monkey named Titan that had been trained to point to the general spot on a map where the *Titanic* sank, which Grimm thought could be silly enough to earn him some free media. He also thought the monkey might break up the monotony of boring scientists reading a bunch of gauges and charts. Once they found the ship, the monkey could be the star of the movie they would make.

One of the scientists, Fred Spiess of the Scripps Institution of Oceanography, thought the monkey was bizarre and would distract from an otherwise serious pursuit. Spiess delivered an ultimatum to Grimm that if the monkey went, the rest of the crew wouldn't.

"Then fire the scientists," Grimm replied, only half joking.

The monkey stayed behind. And when the search ship stopped for a night in Bermuda, Grimm made the odd choice to leave the ship and fly back to Texas. He said there wasn't enough space on board and didn't want to distract from the scientific work. The real reason, however, was that Grimm went back to juice the press of an impending announcement. This was convenient for a man who famously suffered bad bouts of seasickness, and even luckier after the two-week expedition was hampered by bad weather, which made it doubly hard to move in a straight line at a steady speed.

The ship returned to port two weeks later with sonar images of fourteen different sites that appeared like they could be the *Titanic.* The team had fallen short of its goal of making a groundbreaking discovery, but Grimm tried to salvage what he could. Against the wishes of his scientists, Grimm called a small press conference and announced that the sonar pictures would "conclusively prove" that his team had found the *Titanic.*

"I think we got that heifer corralled in a box canyon," he declared.

When he distributed the photos, analysts observed that the images showed only a shadowy blur on the seafloor that could just as easily be rocks. Grimm refused to concede. He said they had found three other ships, or parts of them, and any one of them could be the *Titanic.* The crew declined to confirm that claim, fearing that Grimm's sensational pronouncements could dilute their scientific credibility. The expedition's leader, a man named Mike Harris, simply said in response to Grimm declaring early victory that he remained "highly optimistic."

In private conversations, however, Harris admitted that the crew had been surprised at the "complex and dynamic" terrain of the seafloor. The sonar images revealed that the search area, and the ocean floor in general, were far more variable than anyone anticipated. There were wide stretches of mud and huge outcroppings of rock.

Rather than being flat, the floor sloped in miles-long ridges, inclines, and cliffs. At one point, the sonar showed an object that appeared to be the same width, length, and height of the *Titanic*, which excited the crew. But on closer inspection, it turned out to be the ledge of a sprawling undersea canyon.

Harris and his crew were humbled. But more than any skill, Grimm was perhaps best at keeping people's attention by dropping morsels of allure and new promise, reframing a loss into an inevitable future win. Just as the reporters were packing up to leave, he returned quickly to the podium. He announced that next summer, the summer of 1981, he'd return to the fourteen sites with better cameras and more advanced sonar and prove that he had been right all along.

° ° °

Grimm's scientific credentials were easy to question because he wasn't approaching the *Titanic* as a scientist. The scientists he was funding were happy to be given the resources to perform basic research. But like Charles Smith and Doug Woolley before him, Grimm's plans to find the ship had centered around media and fame and, by extension, him getting paid.

"The big money is to be made in film and photography," he was quoted as saying in April 1980, the same month he partnered with an underwater filmmaker who had won second place at the Atlanta Film Festival for a documentary about atomic bomb testing in the Pacific. As a result, as news of his 1980 expedition began to deflate from an overwhelming success to an inconclusive question mark, Grimm announced plans for a documentary about the 1980 expedition to raise money for the '81 attempt. He lined up Orson Welles to narrate the film and raised $100,000 in television rights. He negotiated a $150,000 book deal for the exclusive story of the search and paid country-western singer Kenny Starr to write and record a custom "Ballad of the *Titanic*."

His money-hungry motivations boiled over one night at the Explorers Club in New York City. Grimm was held in high regard by the club, and not only because of several generous donations. The club had sponsored ambitious expeditions for more than seventy-five years and its flag had flown at both poles, atop the highest mountain peaks, and across every ocean. As a deep-pocketed adventurer, Grimm was welcomed. But during a ceremony where he was to be presented with the club's honorary banner, a reporter stood up and confronted Grimm over whether he even cared about science and exploration or was just prospecting for another easy buck.

"The Explorers Club has always been associated with serious exploratory endeavors," the reporter said. "I wonder, in view of the media hype, second only to the jumping of the Snake River by Evel Knievel, if there isn't perhaps an overriding commercial consideration here that goes beyond the traditional feeling of the membership of the club?"

Before Grimm could answer, the president of the club, an energetic man named Charles Brush, grabbed the microphone and accused the reporter of confusing "commercializing with popularizing." Popularizing, he said, was a crucial part of exploration. It spread the scientific spirit and turned mercurial observations of field researchers into broad transmissions of knowledge. If an explorer made a film and made money off it, all the better, especially if that money fueled future exploration.

What Brush didn't say, however, is that hype for the purpose of money is salesmanship, not science, and when it comes to the world's biggest quests—to find Amelia Earhart's missing plane, for instance—populist zeal can result in scientific corners cut and half-baked assumptions perpetuated as fact.

Grimm's mind was made up on the matter. And he had plenty of support from a new generation of buffs and wingnuts. Prospecting oil had made him rich, but prospecting the *Titanic* had made him famous. Almost every day, a postal worker delivered a wad of letters to Grimm's Abilene office, many from overseas. There were notes from engineers offering mathematical help, correspondence from mediums who claimed they had spoken to victims of the *Titanic*, and requests from people who wanted Grimm to sponsor a search for dinosaurs. A man in Yugoslavia wrote asking if Grimm could help him show his art in Texas, the supposed epicenter of American culture. Grimm answered every letter. He also spent hours on the phone chatting with well-wishers and enthusiasts, all calling to cheer on the man they had read about in their local newspaper. Grimm loved the attention and was known to chat for hours with strangers about the nuances of the wreck, from the frantic telegraph calls to the order in which the lifeboats were loaded. He spent so much time on the phone that he joked that a telephone would one day be installed in his coffin.

Still, for such a flamboyant man, his goals were surprisingly understated. Grimm wanted to be the man to find the *Titanic*, but he didn't care much about the quality of underwater photography, and he certainly didn't want to *salvage* it. A young navy captain named Robert Ballard, who had experience finding and photographing wrecks, was asked in the press to weigh in on Grimm's efforts. He observed that the world would need high-quality images of the ship to prove it was found. People wanted to see the familiar hull they remembered from films and on posters with its unmistakable name across the bow. A blurry sonogram showing possible debris that might actually be rocks wouldn't cut it to command global attention.

For as far as anyone knew, the ship was still in one piece, two at most, sitting on the seafloor, waterlogged and worn down, but still very much recognizable as one of the great ships of all time.

ooo

After sixty-eight years underwater, the wreck was in fact barely hanging on. Of the original ten decks, only four or five likely remained by 1981, according to modern modeling of the breakdown. The ship had been lucky to land on a gently sloping area of seabed swept constantly by the Western Boundary Current, which dusted away sediment that might have buried it. But the lack of sediment also left the wreck vulnerable to constant galvanic corrosion as the interaction with salt water caused the steel to release its electrons and return to its elemental state—iron—which in turn was eaten by deep-sea microbes.

Thomas Edison once experimented with ways to arrest the breakdown of metal on ships. In his day, the dominant problem was that ship metal would weaken in the span of a few years or even months due to constant exposure to salt water. It was clear that the salt pulled electrons from the steel. But working off the math of the earlier British chemist Sir Humphry Davy, he figured that if you could offer the salt water a *different* source of electrons, perhaps from a piece of a weaker metal like zinc or magnesium, it would leave the steel alone, at least for a while. The system became known as cathodic protection, in which a sacrificial cathode metal provides cover to a more preferred metal.

Halting breakdown is most useful on expensive watercrafts like oil platforms and aircraft carriers that require long life-spans to justify their enormous costs. But preserving vessels after they sink can also be helpful for both historical research and environmental protection. The Chuuk Lagoon, once Japan's primary military base in the South Pacific, is one of the biggest ship graveyards on earth. A three-day air assault in 1944 put more than sixty Japanese warships and two hundred aircraft underwater, almost all of them sunk with fuel, oil, and weapons on board. This density of shallow-water wrecks has made the Chuuk Lagoon a Mecca not only for divers but also for people who study how ships break down and ways to protect them.

Every so often, a pair of self-described "corrosion enthusiasts" named Chris and Allison Selman visit the Chuuk Lagoon to place large bricks of zinc around the wrecks, some attached by cable and others directly on the old ships. The zinc bricks are essentially Band-Aids. They absorb the corrosive forces of the salt water to protect the underlying steel. But they need to be replaced fairly often. "The metal wears down over time, so you have to keep adding it, but as long as there's metal there, you can halt the decay of a ship almost indefinitely," Chris told me. Spending one's time this way is extremely kind, at least for fish that might suffer from a sudden release of eighty-year-old diesel. But it's also helpful for the tens of thousands of divers who swim through the wrecks and can see a strikingly well-preserved graveyard of wartime tanks, automobiles, hand-painted porcelain bowls, old motorcycles, radios, torpedoes, bombs, and hundreds of human skulls.

"When we dive the wrecks, we always come up quite emotional," Allison Selman told me one evening from Perth, where she and her husband live. "You can see shoes at the end of the bed, uniforms still folded, people's books. You can imagine how life on these boats just stopped when they sank."

Preserving history is a strong and visceral reason to safeguard ships. Yet the earnest efforts of the Selmans and others who delay the inevitable breakdown of wrecks seemed like Don Quixote levels of both virtue and complete futility, of human energy being wisely spent and foolishly wasted. The line between ambition and delusion is so thin, and in the world of shipwrecks, it sometimes doesn't exist at all.

Over enough time, any human effort is mostly an unfair fight against the forces of nature. Despite the Selmans' efforts, it's easy to see year after year how the wrecks are jostled by typhoons and bleached by the sun. The world doesn't change if a long-ago sailor's shoes disintegrate or some old radios waste away. The Selmans' bigger fear is that the wrecks will start leaking fuel into a biodiverse ecosystem.

The U.S. government shares this assessment. Somewhere in Maryland, the National Oceanic and Atmospheric Administration keeps a list of PPWs, or Potentially Polluting Wrecks. Of more than twenty thousand known wrecks in U.S. waters, several hundred are considered at high risk of oil spills, and a handful of those are expected to release oil imminently, if they aren't already. The worst one may be the SS *Jacob Luckenbach*, a troopship that survived World War II only to collide with another ship and sink in 1953 near San Francisco. It went down with more than four hundred fifty-seven thousand gallons of bunker fuel, some of which has been leaking for decades. In 2002, the California Department of Fish and Wildlife confirmed that several "mystery spills" on California's coastline were linked to the *Luckenbach*.

The *Titanic*'s fuel was coal, not gasoline or oil, which posed little danger of an imminent spill. Yet even if its daily consumption of eight hundred twenty-five tons of coal were diesel instead, there would be few options to delay their eventual release. Had the wreck of the *Titanic* been protected with coats of paint or bricks of zinc, its breakdown might've been slowed, perhaps even stopped for a time. But routinely returning to an aging wreck two miles deep to apply new paint or metal cathodes for the sake of historical posterity or habitat protection would be an investment with diminishing returns. More than half a century after it disappeared, no one knew what to expect about the *Titanic*'s true condition. The bigger goal was just to find it. And on that hunt, no one had ever been closer than Jack Grimm.

° ° °

The second stage of our expedition is pretty much on schedule, we are confident of our success," Grimm said on April 11, 1981. Nine months had passed since his first expedition, with only three to go before his second one. July was the only suitable month to attempt a search expedition in the North Atlantic due to low wind and low swells, and Grimm had waited through the winter and spring for another chance, batting down skeptical questions about whether he could really find the ship.

Once he tallied all the receipts, the price tag for the first expedition had come to $1.25 million. Grimm paid most of the cost, but he believed that the next expedition would be cheaper, to the tune of $750,000, and that if he worked the phones, he could get some sponsors and investors.

As it happened, one practically fell in his lap. Bobby Blanco, a Cuban-born gambler, asked Grimm if he could toss a few nickels into the search. Blanco had hunted for the ship himself and lost his job as a restaurant manager when he spent too long in the North Atlantic the prior year. Blanco had bet all his remaining money on

Grimm. And with one investor secured, more lined up, including Nelson Bunker Hunt, a billionaire who had previously tried to manipulate the market for silver by buying up every ounce he could find, only to see his fortune collapse when the market crashed.

Serious scientists also lined up to join Grimm, mindful that if he found it, their absence might be as notable as their presence. He had again recruited oceanographers Fred Spiess and William Ryan of Columbia's Lamont-Doherty Earth Observatory and an engineer named Tony Boegeman of the Scripps Institution of Oceanography in San Diego. Several more big-name researchers came from Woods Hole in Massachusetts, from where the expedition would depart. For decades, Scripps and Woods Hole enjoyed a mostly friendly rivalry over who was first to make new ocean discoveries.

This sort of unbridled enthusiasm fueled the belief that the second expedition would be the final expedition, and that after the near miss of '80, there was no doubt they'd triumph in '81. The sense of inevitability enticed thousands of people to beg for a spot on the expedition. A third-grade class in Denver wrote to ask if they could tag along. Dozens of outside scientists wrote with detailed plans for how they could help raise the ship. Several young women sent pictures of themselves in bathing suits and offered to do "anything" to join the cruise. More than two dozen psychics said they "knew" where the ship was and offered to point it out—one of whom, an older man, presented as evidence that he had made four round-trip crossings on the *Titanic* before it went down.

These were easy to laugh off. But premature accolades did pose the problem of credit, and who would receive it for finding the ship. Grimm was the obvious candidate, but the scientists were reluctant to perform such extensive work only to see it clobbered in the media as superficial show business. Grimm had lured scientists to the project with a trade—help me find the ship and you can do whatever ocean experiments you want the rest of the time. Some of the younger scientists saw it as an opportunity to study the ocean floor and learn about earthquakes and landslides and the erosion of steel underwater. But Grimm was firm in his negotiation. And with his checkbook, he had the upper hand. He did, however, try to downplay his eventual place in history to keep morale high. He reminded the crew that the German businessman Heinrich Schliemann found Troy in 1873, and the British archaeologist Howard Carter found King Tut's tomb in 1922. Both became famous, but neither eclipsed the bigger news of the discoveries themselves.

Besides, Grimm said, "Wouldn't it be great just to find it?"

Reflecting this anticipation, the '81 expedition crew swelled in size. Grimm arranged for a bigger boat, the *Gyre*, a one-hundred-seventy-four-foot-long research vessel that he rented for exactly two weeks from the U.S. Coast Guard. The *Gyre* was equipped with navigational gear and computers developed by the U.S. space program. It carried four miles of steel cable that connected to a torpedo-shaped device, nicknamed "the fish," that would be towed through the water. Unlike the small sonar rig from the earlier summer, the "fish" carried a more elaborate system developed at Scripps called Deep Tow, which was capable of bouncing sonar signals through ocean sediment in case the ship was partially buried. It also carried sensitive magnetometers and a color camera, which would be helpful when bringing back detailed photographs of the *Titanic*. Grimm, naturally, had planned several steps ahead of this eventuality and arranged to commission the *Aluminaut*, the most advanced submersible on earth, to take humans down to the wreck on a future mission.

The morning the *Gyre* was set to depart, Grimm arrived late, which was a classic power move. His delay ensured the maximum number of reporters were on hand. Rather than talk to them all at once, Grimm, wearing a seersucker jacket and a Texas A&M baseball cap, asked them to line up, and he walked down the line answering questions like a celebrity arriving at an awards show.

"We have highly technical equipment," he told one. "We'll find her."

"The scientists we have are the best in the world," he told another.

"The *Titanic* is lost and we're going to find her."

One young reporter posed a blunt, if impolite, accusation: "You're in this for the money."

Grimm hesitated, then cracked a smile.

"Partly," he admitted.

He boarded the *Gyre* and settled into his cabin. As Grimm unpacked, William Hoffman, the writer on board who had been hired to chronicle the expedition, poked his head in with a more candid question for Grimm. After all the anticipation, the bold promises, and the assumption of success up to and including the bottles of champagne chilling in the ship's icebox, what would happen if they didn't find the ship?

Grimm turned the thought over for a moment.

"I'll probably kill myself," he said.

By the second day, much of the crew was seasick. The chef served fatty pork chops he thought would harden the stomachs of everyone on board. But the waves of the North Atlantic were unmoved by the gesture. Grimm passed from cabin to cabin, gripping banisters, to check on his moaning companions. Barely thirty hours had passed, and the ship's radio had received eight calls from media organizations asking if Grimm had found the *Titanic*.

Grimm agreed with the scientists on a search grid that put the prior summer's fourteen sites at the center of a large rectangle. The area was at the base of the continental slope of Newfoundland, where earlier sonar readings from the U.S. Navy had revealed a complicated landscape. The depths in the rectangle ranged from eleven thousand feet to thirteen thousand feet, and the area was bisected by a marine canyon that was a quarter mile at its skinniest and a mile and a half at its widest. The distance between the canyon floor and the rim was as little as sixty feet in some spots and as much as six hundred in others. Finding a needle in a haystack was too generous a comparison. This was like looking for a penny on the moon.

The uneven terrain underscored the considerable advantage of knowing exactly where the ship sank. If it had gone down a hundred miles farther east, the *Titanic* would have sunk to the continental shelf in little more than five hundred feet of water. If it had sunk a hundred miles farther south, it would be more than three miles deep in a sunken valley. Towing a sonar rig required knowing these variations, but this was based on old data, which didn't account for massive flows of sediment from the Greenland and Labrador currents, which were known to alter the seafloor in a matter of months, if not days.

The reality that both the seafloor and the sea surface were in constant motion meant that it was almost impossible to return to the exact fourteen sites from the summer of '80, and even if they could, and even if they found the ship, it was still likely they wouldn't be able to pinpoint it exactly. The crew traded stories about a Spanish galleon that sank less than a mile offshore from Tampa, Florida, in two hundred feet of water. A salvage crew found it late in the day, so they fixed its position with detailed sight lines from landmarks on shore, but when they returned—and a dozen more times after that—they never found the ship again. Grimm and the crew brimmed with confidence that they'd be able to thread these geographic needles, but anyone remotely familiar with the factors at play had to concede in their quiet moments that there was a high likelihood the *Titanic* might never be found.

° ° °

Even on an exciting mission to the site of the most famous wreck to solve one of the biggest mysteries, the overwhelming sensation aboard the *Gyre* was boredom. It would take at least three days to cross half of the Atlantic and arrive at the site. Making matters worse, a storm hit the *Gyre* east of New York. Wind gusts and bursts of water kept everyone in the galley with little to do but eat, sleep, and play cards.

Grimm rationalized the cost of the expedition on the video content he would gather. That meant that as long as there was no discovery, nothing was too mundane to be filmed.

"To make it true to life, we should show you winning," someone said to Grimm as he and a group of scientists played a game of five-card draw.

"That's all right," said Grimm. "It'll be true enough."

"I think Grimm has four of a kind," someone else said during one hand.

"Do you know what the odds are against that?" another player asked.

"Odds don't apply to this character."

During more downtime, Grimm, ever the gambler, suggested the thirty-member crew create an "informal lottery" on which of the fourteen sites would be the discovery site. Grimm divided a map of the search area in grid lines and told everyone to put their name in a square.

One of the Scripps scientists picked site 13, where images from the earlier summer showed a linear object about the length of a ship.

A cameraman selected number 10, where three objects cast shadows along the wall of the canyon.

Mary Linzer from Scripps, one of the few female scientists on board, picked site 7, where a possible field of debris appeared to measure eight hundred by two hundred fifty feet, an extremely rough approximation of the ship.

When the rest of the scientists refused to play on account of not wanting to "root" for a specific outcome, Grimm upped the stakes. Whoever picked the winning square, he said, would win a weekend in Las Vegas on his dime.

Before arriving at the search rectangle, no member of the science crew wanted to be on night watch. But once the search began, virtually everyone volunteered to stay awake all night in the ship's lab, where news of the discovery would be confirmed first.

This made it imperative that the film crew be awake around the clock and ready for "the moment," as though an assumption made from grainy photographs would be immediately followed by balloons and confetti. During the day this meant capturing scenes of waves crashing over the boat, "the fish" being released into the water, and then being pulled up again. At night they trained their cameras on the sonar monitors and the steely gaze of whoever was looking at them, usually Grimm or the head scientist, Fred Spiess.

On day three, when the "fish" was fully extended and dragged over site 6, a spot that the prior year had shown a ship-shaped object, the magnetometer did nothing.

The same thing happened a few hours later over site 13, the object that cast a rectangular shadow.

No one was surprised when site 7 was a dud, but everyone felt the disappointment when site 1, Grimm's pick, registered a flat line on the instruments.

"Like a dead man's heartbeat," someone put it.

Sites 6 and 10 were ruled out too, which rounded out the list of high-priority sites.

People were sick with disappointment, but mostly, they were just sick. The storm from New York hadn't flagged for several days and the ferocious rocking of the boat and the Niagara Falls–level gushers of water left everyone wishing they were back on land. Reports came over the radio of nearby ships that had lost people overboard. Twice the *Gyre*'s captain, a man named Don Armand, considered abandoning the search to go help the other vessels in need. Everyone knew that the North Atlantic was a historically deadly shipping route, underscored most vividly by the folklore of the *Titanic*. But few truly understood how dramatic it still was. In February 1982, several months after the *Gyre*'s expedition and not far from where they were searching, the world's biggest oil rig, the *Ocean Ranger*, capsized, killing eighty-four. A day and a half later, a Russian cargo ship, the *Mekhanik Tarasov*, sank too, killing thirty-six.

By day six, the earlier lighthearted camaraderie on board began to transform into a sense of fatalism and disappointment. Hardened stomachs had turned the mood sour. The camera crew began to blame the scientists for wasting time, and the scientists grumbled that Grimm's paparazzi operation was a circus. The initial symbiosis of science and entertainment had dissolved into resentful attitudes that the "other" side was the reason for the lack of progress.

Two days later, cliques began to form at different tables during the meals. People slept in and missed breakfast. While the control room had been packed during the drags over earlier sites, hardly anyone showed up for sites 11 and 12. During the launch of an inflatable zodiac boat to try to get some underwater footage, one of the scientists misjudged her water entry and almost got her leg chopped off by the boat's motor. Then a part of the "fish" broke and had to be whittled back into place. Nothing seemed to be going right. They were also running out of time.

Scanning the ocean floor didn't turn up nothing. Photographs taken during a long run through a canyon basin—an area that a Canadian marine geologist would later name *Titanic* Canyon—showed a cannon barrel, an air duct, a length of pipe, and a piece of metal the size of a king bed. But even if one assumed these were bread crumbs leading to the *Titanic* itself, there was no way to prove it. The cannon and the pipe could have been from any of the hundreds of ships that sank over thousands of years over one of the world's most trafficked maritime routes. Searching an empty desert would be easy. The bottom of the sea was full of so much junk from so many eras that you were just as likely to find a Viking ax as a color TV.

Grimm's crew was running into a familiar dilemma for wreck hunters who start their search hoping to find any sign of human life and are quickly overwhelmed by how much human garbage is down there. There are artifacts like old carvings, jewelry, or ceramic china. But there are far more aluminum cans, fuel containers, or broken bottles that no one would ever want. Today, a Japanese agency keeps a detailed database of underwater debris, categorizing every tire, soup can, and wine bottle it has found underwater all over the planet. But that's only what's visible; most isn't. A debris-tracking program run by NOAA and the University of Georgia estimates more than eight million tons of trash get added to the oceans every year. And that's a low estimate. Three studies in 2015 calculated there were more than five trillion pieces of debris in the ocean. Some of it floats, but eventually, it sinks, which explains how there could be four billion man-made objects *per square kilometer* in the deep sea, most of it single-use items like soda bottles or plastic grocery bags, both of which have been observed in the deepest parts of the seven-mile-deep Mariana Trench.

How does the equivalent of fourteen million elephants' worth of junk get into the seas every year? In 2018, a group of South African researchers devised an experiment to find out. They went to the aptly named Inaccessible Island in the South Atlantic between South Africa and Argentina, an island covered with ocean trash. For three months, they examined the items for signs or origins or date stamps. They found that the number of the most common ocean throwaways had increased more than 15 percent *every year* over the last thirty-five years—and a striking amount from the prior two years. Most items bore Chinese writing. But this was strange, since even the most hydrodynamic bottles would still take about three years to drift from China to a small island on the other side of the world. They scratched their heads and arrived at the only plausible culprit: Chinese merchant ships, whose activity in the South Atlantic had increased since the 1980s at almost the same rate as the build-up of trash on Inaccessible Island. "It's inescapable that it's from ships," Peter Ryan, the lead researcher, told the *Agence France-Presse*.

Chinese ships aren't the only offenders. Until the 1970s, almost any ship in international waters could dump whatever it wanted. Most availed themselves of the convenience under the legal principle of "who's going to know?" Newer regulations set dumping parameters based on distance from shore, at least in U.S. coastal waters.

Once a ship is more than three miles from port, it can dump small wads of paper, cardboard, or food waste. More than twelve miles from shore it can let loose metal and glass and almost any trash, so long as it's weighted to sink to the bottom. Plastic is verboten everywhere, and sewage—so long as it's treated—is allowed almost anywhere. One imagines that while Congress deliberated this final provision, a humble aide stood up to make the irrefutable point that if whales, sharks, and dolphins are allowed to poop in the ocean, why can't we?

As time went on, the most damaging effects of ships turned out to be not from human waste but from manufactured goods that cross oceans on giant container ships. The burst of twenty-first-century globalization had the predictable effect of making ships bigger. The vessels that brought fruit from Ecuador or flat-screens from China grew to accommodate more capacity, at least to a point. Most container ships on earth have been designed to squeeze through the one-hundred-ten-foot-wide locks of the Panama Canal. But the construction of a wider one-hundred-eighty-foot canal lane in 2016 and plans for an even bigger canal across Nicaragua fueled a shipbuilding boom.

Bigger ships suggest fewer trips across oceans. But the opposite happened. Bigger ships allowed for more than double the quantity of goods crossing the seas by 2015. The insatiable appetite for manufactured goods—fed largely by e-commerce—fuels projections that ocean freight will triple in the Atlantic and quadruple in the Pacific and Indian Oceans by 2050.

The unseen effect of ships loaded with millions of pounds of freight is that they can be easily knocked off course or off-balance. The general public rarely sees this, but in early 2021, a cargo ship a quarter mile long called the *Ever Given*—a ship visible from space—

made international news when it lodged its bow into the banks of the Suez Canal. Its momentum from carrying twenty thousand shipping containers swung its stern to the opposite bank and blocked one of the world's busiest shipping bottlenecks for six days as dozens of tug boats, excavators, and frustrated Egyptians worked to refloat it. Around the same time, the *Maersk Essen*, a cargo ship carrying thirteen thousand shipping containers, rocked so forcefully on its way from China to Los Angeles that it dropped seven hundred fifty containers into the Pacific Ocean. Several floated, but most sank and will probably never be found, along with thousands more containers lost at sea every year. Ocean trash is bad enough. But it's a bigger shame that there are millions of pairs of new sneakers, cheap umbrellas, and multicolored beach chairs that no mollusk, starfish, or sea snail will ever want or ever use.

∘ ∘ ∘

Jack Grimm wasn't looking for debris, human or otherwise, and he wouldn't be satisfied if he found it. After the *Gyre* had towed the "fish" over the last of the fourteen sites and each one had been photographed, studied, and ruled out, Grimm said it was time for the grand finale. He ordered the crew to pull the "fish" out of the water and replace it with waterproof film cameras. If they weren't going to find the *Titanic* using the magnetometer, at least they could take a few hours of footage of the seafloor that could be analyzed later for signs of the wreck.

This was every bit the Hail Mary it sounded like, and the strangest thing happened: it worked.

Later that night, as the *Gyre* and its defeated crew headed back to Boston, the television in the lounge showed the tape from earlier that day. Everyone gathered around to gawk at a starfish as big as a car tire and a shrimp longer than a foot. Fish swam in front of the lens as if posing for a portrait. There was an old bottle, a tin cup, and something that looked like the skull of George Washington.

And then in the final thirty minutes, the seafloor changed suddenly in texture from sandy nothingness to a distinct object.

Grimm sprang out of his seat as if shot from a gun.

"Wait a minute!" he yelled. "Play that back!"

Chapter 11

# ALL THESE MOTHS DRAWN TO THE SAME FLAME

The propellers built for the *Titanic* and its sister ships had been among the largest in the world at the time. At twenty-three feet across, they ran on a triple-screw system, which gave the *Titanic*, the *Britannic*, and the *Olympic* three propellers each—two side by side and a smaller one in the middle. Exactly how big to make each propeller was a complex calculation of fluid dynamics to discern how fast water would flow through the propellers. Then came the matter of pitch, or how angled they should be. Too steep and you'd get no propulsion at all; too sloped and the ship would be pushed up instead of forward.

The *Titanic*'s propellers were cast from manganese bronze, and their design had been inspired by an 1836 patent by the British inventor Francis Pettit Smith, who designed a propeller like a screw. The thread extended two full revolutions, like a spiral staircase, which Smith believed offered twice as much propulsion than would a single revolution. At a demonstration on a small boat at the Royal Adelaide Gallery of Practical Science in London, Smith explained that the device transferred rotational motion into linear thrust by creating a pressure differential in the water, thus moving a ship (and later an airplane) forward. The bigger the propeller and the quicker it turned, the greater the propulsion.

Smith's audience was easily persuaded. Within three years, the Royal Navy committed to outfitting all future ships with screw propellers, starting in 1838 with the steamship the SS *Archimedes*, named after the ancient Greek mathematician who invented the screw. There was no question the screw propeller was superior in battle. Compared to a paddle wheel, which could be attacked and destroyed, a propeller was underwater, which also allowed for more guns and cannons on the stern of a ship.

Yet the question of whether a propeller was more *powerful* than a paddle wheel could be settled only one way: a tug-of-war. In 1845, navy sailors tied a rope between two comparable ships—the propeller-driven HMS *Rattler* and the strong paddle steamer HMS *Alecto*. Then they accelerated both vessels in opposite directions. When the *Rattler* pulled the *Alecto* backward at more than two knots, the matter was settled. Paddle wheels were demoted to the slow ships of history, like riverboats and lake ferries.

The *Titanic*'s propellers, however, reflected modern breakthroughs in ship speed. No, it turned out, a bigger and faster-turning propeller wasn't always the most powerful. Large blades provide thrust, but also extra drag. Three blades were more powerful than four, but four provided balance and kept the thrust more even. This combination resulted in the *Titanic*'s sleek and optimized design of the three propellers—the pitch of each blade nearly equal to its diameter—and made from expensive bronze instead of steel to keep the smooth, sleek blade from rusting. The two outer propellers had three blades and the smaller inner propeller is believed to have had four blades, although this has become an odd point of controversy in the *Titanic* community. For all the expense, weight, and size of the propulsion system of the world's most famous ship, not a single pre-sinking photograph has been known to exist of the propellers, fueling endless debates about the number of blades on the sole center propeller.

At least until Jack Grimm thought he spotted it.

"Play it again!" Grimm kept saying. "Go back!"

The cabin was quiet as everyone crowded around the monitor, faces almost touching the glass, as the five seconds of tape were looped over and over. The endless expanse of textured mud was interrupted by a large object that looked like a butterfly wing, large and flat and coming together on one end to attach narrowly to a piston. It was too curved and sleek to be a rock, and it was also enormous, looking every bit the twenty-six tons of a *Titanic* propeller blade.

Grimm felt in his bones that he had encountered the underbelly of the ship that had fascinated and eluded him for years. And based on this five-second video clip alone, he was convinced. No one in the control room dared to contradict him or wonder aloud if the object could be something else, or perhaps belong to one of thousands of different ships that had ever wrecked or lost a propeller in the North Atlantic.

Grimm called for Captain Armand, who was almost as eager to find the ship as Grimm. No one was to say anything to Armand before he saw the footage.

"It's a propeller," Armand said immediately.

"From the *Titanic*?" Grimm asked.

"It's consistent in size."

Grimm made a beeline for the radio room. The Coast Guard was expecting the *Gyre* back immediately, but surely they'd grant an extension considering the circumstances. Talking to a Coast Guard officer on the other end of the line, Grimm first turned on the charm (*Aw shucks, you wouldn't believe it!*). Then he moved to reason (*Now listen, if we leave now, we'll never find it again*). And then to name-dropping and political clout (*President Reagan is a friend of mine, and I know he'd want us to get this done*). When it all failed, he resorted to begging, and when that too failed, he got angry and fell to insults and threats. Finally, he flung the radio receiver across the room.

Even if Grimm could convince Captain Armand to buck the authorities and swing the helm around, risking both men's finances, reputations, and potential liability, it would have been extraordinarily difficult to navigate back to the precise spot of the propeller. But it was moot. Armand knew they couldn't turn back. After the *Gyre* returned to port in Boston and the thrill faded, the crew disembarked with wistful nostalgia and promises to stay in touch.

Not Grimm, though. Grimm sat alone stewing. He was angry and tired, feeling he had come so close and still left empty-handed. But his eyes were fixed firmly on the future, obsessed with the image of the possible propeller and what else might be nearby. A man not prone to giving up until he was at the end of his rope, and then a mile or two past that, he announced to reporters on the dock that he had "definitely" discovered the *Titanic* and, in eleven months, would go back for the third time and bring back detailed photos.

Grimm's claim of his discovery made international news. But without irrefutable proof that the propeller was in fact a propeller, and more specifically the *Titanic*'s propeller, the story stalled. As

Grimm's photos made the rounds, marine archaeologists generally dismissed the object as a propeller-shaped rock outcrop.

That turned out to be true. But what was also true was that Grimm had come closer than anyone knew, including himself. Years later—and only with the benefit of knowing exactly where and how the ship had fallen—Grimm was made aware that on his fourteenth and final target, the "fish" passed just five hundred feet from the bow.

○ ○ ○

How might it have felt? What might have happened if Grimm had towed his rig directly over the *Titanic* and brought up the first color images the world had ever seen? For years, Grimm had fixated on the reveal phase: the clothes he would wear as he was carried off the *Gyre* like a king, the speech he'd give at the press conference announcing to the world he had done the impossible. But what would it have felt like, deep down, in those early moments when it set in that the quest of many men's lifetimes had been solved by *him*? Most people are happy simply to find their lost glasses or misplaced keys. But for men like Grimm, once you made a sport of finding missing items, the only way to scratch the itch was to search for things bigger and bigger.

I had been warned for months that nobody talks, writes, or even thinks about shipwrecks for long without coming across David Mearns. Shipwreck hunters are an abundant species, but shipwreck *finders* are a more elite club, one you can't talk your way into with bravado and deep pockets. Owning salvage rights to a ship or chartering a boat to comb a search area are small maneuvers compared to the triumph of declaring a riddle solved.

In 1990, Mearns's first find was his biggest. His discovery also helped solve a crime. The *Lucona* was a Panama-registered freighter that sank in the Indian Ocean in 1977. Udo Proksch, the Austrian man who owned the cargo, said the ship had been carrying "expensive uranium mining equipment" and filed a hefty insurance claim for $20 million. With little proof, the insurance company paid the claim, but thirteen years later, when Mearns found the *Lucona* in almost fourteen thousand feet of water, he not only couldn't find the mining equipment, but he also noticed the remnants of a time-release bomb, suggesting Proksch faked the accident at considerable cost, including the lives of the six men who died in the sinking. A handful of Austrian government officials were discovered to have been in on the plot, signing off on the fake cargo and obstructing an investigation into the incident. Proksch escaped to the Philippines in 1988 but returned to Austria a year later, when he was recognized in a disguise and arrested. Evidence from Mearns's discovery of the *Lucona* landed Proksch a prison sentence of twenty years.

Mearns could hang his hat on the *Lucona* alone, dining out and giving speeches about how he dug through historical records and plotted points on maps. But he felt drawn toward bigger mysteries and deeper wrecks. In 1994, Mearns found the MV *Derbyshire*, the biggest British ship ever lost at sea. It was nearly three times the displacement weight and sat at almost the exact same depth as the *Titanic*. Several years later, he found the British two-mile-deep battle cruiser the HMS *Hood*, which had been sunk by a German battleship in 1941. Like the *Titanic*, the *Hood* had also been a symbol of invincibility that fueled decades of enthusiasm and fascination. Every expedition for Mearns seems to bring a new announcement of a new discovery: in 2008, the Australian battlecruiser the HMAS *Sydney*; in 2015, the Japanese battleship *Musashi*; and of course the Guinness World Record for the deepest shipwreck ever found, the German runner *Rio Grande*, waterlogged under three and a half miles of water.

I finally got ahold of Mearns one morning. Slim, with gray hair and a goatee, he looked more like a mid-level accountant than a grizzled ocean veteran. What was his secret, I wondered, that allowed him to turn up dozens of wrecks that other people—many of them extremely smart, experienced, and well funded—had tried to find and to make it look almost effortless? It was clear from the stories of Woolley and Grimm that obsession wasn't nearly enough.

"You'd be surprised how much you can learn in books and old newspapers," he said. "But the most important part are eyewitness accounts."

"Like accounts from survivors?" I asked.

"Yeah, and people on other ships in the area."

He explained that, like Jack Grimm, he had found details about the weather and the water current. Was there wind that night? More than once he tracked down descendants of survivors—people who weren't even alive when the ship sank—and they said something like, "My dad used to say it was a full moon that night," which led Mearns to a new breakthrough. Often, even harder than finding the ship's location was raising the money for each expedition. Investors weren't interested in old battleship radios or navigation computers. They wanted flashy finds like gold or diamonds. He often partnered with TV companies or foreign governments with a media or historical interest. Occasionally philanthropists got involved, but only if a big anniversary of a battle was coming up, thereby guaranteeing free media and public credit that, naturally, could be monetized.

Mearns was good at what he did, but even for the best hunters, there were still ships out of reach. For a long time, the one that haunted him was the *Endurance*, the ship Ernest Shackleton took in 1914 to

Antarctica, where it was sandwiched by converging ice floes and spawned one of the greatest survival stories in the history of exploration. In 2003, Mearns met Shackleton's daughter, who gave her family's blessing for him to go find the *Endurance* and recover its artifacts. The only problem was that the *Endurance* was almost two miles deep at the bottom of the ice-covered Weddell Sea, an area almost as dangerous to ships today as it was in Shackleton's era. I understood the challenge facing Mearns or anyone who looked for the *Endurance*. But then, as this book was heading to the printer, the darndest thing happened: someone found it. A team led by the polar geographer John Shears and the maritime archaeologist Mensun Bound—a man sometimes called the "Indiana Jones of the deep"—used battery-powered drones to comb 150 square miles of Antarctic seabed 10,000 feet deep. The drones worked for two weeks until they returned with high-definition photos of the *Endurance*'s bow and its helm. In the wreckhead community, this was enormous news, a discovery almost as seismic as finding King Tut's tomb or landing on the moon. But for the biggest fans—and competing hunters like Mearns—the news also seemed to have an undercurrent of disappointment. During its century-long absence, the mystery of the *Endurance* had become more interesting than the *Endurance* itself. Now that it was revealed to be exactly where everyone thought it might be, looking exactly like everyone hoped it would, there wasn't much left for anyone following at home to do but pack up and move on.

"How come you never went after the *Titanic*?" I asked Mearns.

The timing was off, he said—he didn't start searching for wrecks until the late eighties, when the ship had already been found. If it was still undiscovered by the mid-nineties, he might have been the one to find it. And he would have had little trouble assembling investors after his early success with the *Lucona*, a badge of credibility that Doug

Woolley and Jack Grimm lacked. I knew that in 1992, seven years after the *Titanic* was found, Mearns and his company, Blue Water Recoveries, were hired by investors to salvage objects from the *Titanic*'s debris field. But I didn't know that one of those investors was Jack Grimm, who, in the years after someone else beat him to the wreck, had turned to collecting artifacts at the site that could be sold, licensed, or profitably put on display. As it happened, Grimm's hairline had nowhere left to recede on September 30, 1992, when a federal judge in Norfolk, Virginia, ordered Grimm and Mearns—whose boats were idling at the wreck site at a cost of $1,000 per hour—to keep their hands off the debris until he could sort out who the artifacts actually belonged to. This took months, and the men eventually had to go home. To this day, Mearns still blames the judge for screwing them up. But he also learned a valuable lesson: with shipwrecks, the only thing more important than good research is a good lawyer.

I could sense Mearns had long ago grown tired of talking about the *Titanic*, the same way an aging musician grows weary of playing his early hits over and over. It wasn't even *his* wreck, and he couldn't escape it. He didn't seem to mind that it was cliché, its originality stripped by decades of cultural dissection. What bothered him was that it was crowded—an endless field of wackos circled it like vultures competing for every last scrap.

"You have all these moths drawn to the same flame," he said, clearly frustrated. "And what happens when a ship gets that kind of attention is you get all sorts of people who have no business dealing with the *Titanic* claiming that part of the ship is theirs. It's ridiculous."

Mearns seemed to be pointing at a distinct specimen I had come to know well. Charles Smith, Doug Woolley, and Jack Grimm had all been drawn to the shiniest of shipwrecks. But their lack of experience finding or even looking for lost ships had posed repeated obstacles, first in their difficulty raising money for expensive search expeditions and then in their poor estimation of how to get a collapsed passenger ship off the floor of the Atlantic. All were forced to rely on the expertise of others and to build teams of scientists if they could afford it. And if they couldn't, to simply declare they had. Surrounding oneself with smart people can work if you're the president or a CEO, but for a technical job that no one cared about as much as them, knowledge and experience were the missing link.

As it happened, around the same time Mearns was schooling me about *Titanic* hangers-on and how to find lost things, I happened upon the clue that led me to Doug Woolley himself, alive and in the flesh. The newspaper article, his home address, a series of phone calls, and a flight to London. It had been a long time since a journalist had asked to hear Woolley's story, and here I was, unknowingly about to rekindle the hope of a long-ago dream.

---

Grimm kept his word and elbowed his way through another round of investment for another expedition in the summer of 1983. Compared to the sensationalism of his first two attempts, Grimm approached the mission with a sense of reality, even fatalism. He had already spent more than $2 million on the multiyear hunt, and as desperate as he was to find it, he hadn't made his oil fortune by repeatedly drilling the same dry well.

"Third time's the charm," Grimm said several days before embarking. "I'm not going back if I don't find it this time."

In addition to his own money, Grimm had exhausted the resources of his rich friends. The grant money that was free-flowing for the first and second attempts had slowed to a trickle. After the second expedition in 1981, Grimm commissioned the publishing of his own book, *Beyond Reach: The Search for the Titanic*, which was both about him and by him, written—similar to Woolley's book—in the third person.

When the book came out, Grimm thought of a Willy Wonka–style gimmick to drive book sales. He announced that somewhere in the world, he had hidden an eighteen-inch-long miniature replica of the *Titanic* made of seventy ounces of gold worth $25,000. He imagined that any mention of the public prize would have to mention the book title, which would juice sales to cover the cost of the replica *and* another expedition. But a year later, two men found a photo of the golden ship that Grimm had placed in an obscure book in the New York Public Library. They demanded cash instead of the gold, which Grimm begrudgingly paid. Meanwhile, the book sold a disappointing fifteen thousand copies.

The third expedition didn't go much better. Aboard the *R. D. Conrad*, a new research ship operated by Columbia University, Grimm made a beeline for the propeller spot. But bad storms ended the effort almost as soon as it began. Six days later, the crew, "half sick and cross eyed," returned to port with barely more than new blurry sonar images of three new targets that Grimm said looked suspiciously like the *Titanic*'s hull. Storms prevented a closer look.

Publicly, Grimm pulled from his familiar playbook. "We are all very pleased," he said. And he mused about a fourth expedition to go back. But privately, Grimm seemed to accept defeat. Finding the ship required more time and more money, and he was short on both. He hated being on the water almost as much as he hated burning through his millions. More than both, however, a man who spent his whole life believing he was a winner had had enough of losing.

There's probably no better place to ponder the merciless nature of the Atlantic than on the Atlantic itself. But since I wasn't about to take a boat from California to London to meet Doug Woolley, the best I could do was peer down from above.

The Atlantic likely doesn't have as many wrecks as the much-bigger Pacific, but it does have an impressive trophy collection. The Atlantic has some of the oldest shipwrecks ever recorded, including a Bronze Age ship from the fourteenth century BCE. It holds the most ancient wrecks known to sink in the same year (two hundred forty ships in 480 BCE). And it has one of the biggest warships, the German battleship *Bismarck*, which sank in the North Sea in 1941. The turbulent waters off North Carolina are known as the "graveyard of the Atlantic," a title that's also used to describe areas off Nova Scotia and Boston. Around every American port, in New York, Virginia, South Carolina, and Florida, the waters in spots are so dense with wrecks that if you were to empty the ocean, a person could spend a day wandering around looking at dozens of them. A precise number is elusive, but one researcher I spoke to suspected that based on the arc of history, more people had died on boats than ever will in car accidents. "We make it easy to forget how dangerous it is, but in the past, they knew," he told me.

I stared out the window on that flight imagining what it must have felt like to be Columbus or Magellan, or worse, one of their hundreds of crewmen, who faced the extreme likelihood that boarding a boat and sailing over the horizon might be the last thing they ever did. Leaving your family and country to go somewhere new seemed exciting, until day three, when the endless bobbing and no land in sight emptied even the most durable stomachs. I imagined going months without seeing a fresh fruit or vegetable and feeling my scurvied body collapse in on itself, all while surrounded by the squalor of rodents and the incessant abuse from the most impatient and often drunkest men alive. William Golding's 1954 novel *Lord of the Flies* offered a fictional horror of what would happen to a group of young men left on an island alone. But few who read the book probably paused to consider that the same collapse of governance, decorum, and basic humanity had occurred thousands of times on ships in remote waters everywhere. Tension would grow until it exceeded the bounds of the ship, perhaps illustrated best when British captain William Bligh faced his famous ejection from his ship the *Bounty* in 1789 in the middle of the South Pacific. Like many exiled captains and crewmen before him, he would have died, except he was lucky that several loyalists rowed him thirty-five hundred miles to safety—longer than the distance from California to Florida—in a primitive rowboat.

The absence of such stories today reflects two centuries of breakneck ingenuity and maritime innovation. Ships continue to wreck, and occasionally people wash up on deserted islands. But they're often anomalies and never become marquee cultural events. The best evidence might be the thirty million people each year who sign up for cruises, thinking of seafaring as the ultimate way to *relax*. Or that when someone gets bored or rich—or both—the obvious thing to do is buy a boat.

Imagining a different reality without luxury yachts and airplanes, a world one or two hundred years ago, when I'd have had to take a steamship to London or row there in a skiff, made me wonder about a different parallel reality. Looking down at the waters that held the *Titanic*, I tried to imagine what it would've taken to truly raise the *Titanic* from the ocean floor. To return the ship to the surface and, as so many people in the twentieth century had envisioned, return it to glory as a museum or hotel or, in the wildest fantasy, recommission it for renewed pleasure cruising.

Over the course of writing this book, I had become obsessed with this thought experiment, the question of what it would take—scientifically and financially—to do what people for decades had thought was so easy.

Every time I read something or talked to someone about the chemical degradation of a century-old wreck and the physical limits of working two miles deep, I eventually steered the conversation and asked them to dream.

Say that it was possible, I would say. How much money would it take, and how would you do it?

I asked this question dozens of times, and David Mearns was the first to refuse to indulge me. It wasn't possible, he said, and even if it was, there are hundreds of other ships more worthy of such a gargantuan engineering effort.

"Realistically it can't be done, even with all the money in the world," he said. The problem wasn't just depth and ocean pressure. It was that the ship had no coherent structure. He likened it to the Twin Towers after they fell. How would you pick them up? What would you even be picking up? And that's before adding more than twelve thousand feet of water. It was a nonstarter from fantasyland, he said. "If people thought they could do it, it's because they didn't know what they were talking about."

I couldn't argue with Mearns, a man who had been to the other side and seen the deep-ocean landscapes and decaying debris that the rest of us couldn't conceive of. But I wasn't willing to give up the notion that, at some level, it *was* possible. If raising the ship's hull and its debris field had been a national priority, like the *Challenger* space shuttle, or had the urgency of a national security mission, like the *Hughes Glomar Explorer*, the U.S. government, thousands of scientists, and the strongest military on earth could make it happen. These were the powerful forces that figured out how to split an atom, map a genome, build the internet, and walk on the moon.

The more I asked, the more colorful ideas I got. Some people said that they'd build a giant cradle with an autonomous propulsion system to visit the site and envelop the wreck in a hydraulic claw. Another offered a more tedious but easier approach: for a team of autonomous underwater vehicles to work continuously for months retrieving every ounce of debris, categorizing them in a master geospatial database to be put back together on land.

Other people said that you didn't need all the *Titanic* debris to claim success; a triumphant recovery effort could just get a few big pieces to go on display in museums. A New York–based team actually tried this in 1996. The crew spent almost an entire summer trying to lift an eleven-ton section of the ship's hull—a piece of steel twenty-four feet long and twenty feet wide with four portholes—by attaching four rubber bags filled with diesel, a liquid *slightly* lighter and thus more buoyant than water. Surprising everyone (and probably themselves), they got the piece within two hundred feet of the surface, but while towing it to shallower water, the piece broke loose and sank again. Charles Pellegrino, a scientist and consultant on the project, blamed supernatural forces. "I don't believe in the curses [of the *Titanic*], but I don't *dis*believe in them at this point," he said. Two years later, a different crew went back and got the piece. It was treated with acid-resistant derusting spray and found its way into a *Titanic* exhibit in the Luxor hotel in Las Vegas.

The closest I found to a credible idea to raise the entire ship came from a man who was once a young engineer at the U.S. Naval

Academy and had studied the prospect in detail. In 1977, the film company making *Raise the Titanic* commissioned the Naval Academy to come up with a technically plausible scenario that could be written into the script and portrayed with cinematic believability.

The young engineer, Parks Stephenson (whom you might remember from chapter 3), was a student at the time and offered to assist his professor, a naval architect named Roger Compton, with the inquiry. Stephenson spent weeks poring over books on maritime architecture and ocean physics. He studied cross sections of the decks and performed lengthy calculations on drag, lift, and buoyant force. He accounted for the "calcareous ooze" and "pelagic mud" of the seafloor and its suction force on an aging wreck.

In the final report, Stephenson acknowledged that his analysis had major caveats. He assumed the ship had sunk in one piece and sat upright on the seafloor, which in the absence of evidence remained conventional wisdom in 1977. Still, this was for Hollywood, not the Army Corps of Engineers, so facts could be fudged.

The first idea Stephenson and Compton came up with bore a stunning resemblance to Charles Smith's initial plan in 1914—namely, to apply enough buoyant force to the hull to rip it from the mud, then pull it up successive levels as it was towed to port. It would only break the water surface when it was close to shore, at which point divers would attach ropes and anchors to the hull and then, presumably, a series of cranes on land would pull the whole thing out of the water, dripping for days as it emptied its bowels.

This, however, was unacceptable to the filmmakers. No one would go see a movie about a ship towing another ship *underwater.* They asked Stephenson for a more dramatic plan B that would eject the ship out of the water like a breaching whale. Stephenson did more calculations and then submitted the simplest plan he could: fill the hull with syntactic foam full of tiny little air bubbles, a material not much different than millions of tiny Ping-Pong balls. Put enough in and the ship would shoot upward like a missile until it triumphantly returned to the surface of the Atlantic.

The filmmakers liked this idea, ignoring the question of how you'd evenly and quickly inject millions of pounds of syntactic foam underwater. When it came time to film the scene, the filmmakers built a fifty-foot-long scale replica of the ship. It cost $7 million—roughly the same price tag of the original *Titanic* in 1912. The filmmakers couldn't find a water tank or swimming pool big enough for the scene, so they built a ten-million-gallon "horizon tank" on the island of Malta that would line up with the actual ocean and look on-screen like an endless sea.

People who make movies know that when filming sinking scenes, or any scene where something pretty gets ruined, you get one shot. But there's no limit to the reverse: taking something ugly and decrepit and making it clean again. To film the climax scene of *Raise the Titanic*, the ship was lowered and raised more than fifty times before the crew got the shot they wanted. Down and up, wet and dry, sinking and unsinking, the worst day any ship could possibly have, with the added indignity of being surrounded by cameras.

There's an odd catharsis watching a ship surface under its own power, like an injured racehorse that stands up and runs again. This final four-minute scene was about as realistic as a dog doing calculus or humans landing on Pluto. But perhaps because of that, I was glued to it for days. After dinner or first thing in the morning, I'd return to my computer, replay the clip, and watch the *Titanic* rise again. I figured out how to make the slow-mo video even *slower* and watched the bubbling whitewash pour off the bow and out of each porthole. I marveled as the camera panned the decks covered in mud and how the railings were mangled by time, like worn-down paper clips. Along with the violin-led score, the movie magic of a shipwreck unwrecking itself offered a rare dose of redemption in a field so filled with tragedy and destruction.

Who wouldn't agree with that? That a fake raising of the *Titanic* was better than none at all, and that getting to see it on-screen gave at least a *replica* of the *Titanic* another shot.

But that would require unanimous agreement that the time for *Titanic* rescue missions had indeed passed—that there was nothing left to do with the old ship except visit its deep grave, study its breakdown, and pick up any remaining loose ends. You could probably ask one thousand people and all one thousand would agree that it was time to let go of the pipe dream of raising a broken ship and move on.

But if you asked a million, or perhaps a billion, odds were you'd get at least one person who still believed the impossible was possible.

And if you found him, as I did, living in East London, he would explain that it wasn't him who was delusional, naïve, and dimwitted. It was you.

Chapter 12

# MAN IS NEVER LOST AT SEA

Doug Woolley sent an assistant to meet me at Heathrow. He made it sound like I should expect one of those men in black suits who holds a sign for business travelers. Instead, I was greeted by a young fellow, about my age, named Kamran. He was tall and frazzled. He was also confused as to who I was or what he was supposed to do with me. We shook hands and chatted for a moment. I asked where he parked.

"Oh, I took the train, mate," he said.

We got on the tube and for the next two hours, we made small talk about the weather and the reliability of public transportation. Every time I steered the conversation to Woolley and benign questions of his past, Kamran told me I'd have to ask the man himself and changed the subject.

I knew Woolley would be gregarious. I had seen old videos of him explaining how he owned the *Titanic* and what he planned to do with it. I watched him dress down a confused British news anchor who was unsure whether to take Woolley literally or seriously, both or neither. I also learned from Kamran that Woolley had been looking forward to my visit for weeks.

"I don't think an American has ever come to see him," Kamran said, an assertion that left me wondering if I was savvy to make the trip or foolish. Later, another friend of Woolley's told me that he considered my visit the highlight of his year.

It occurred to me later that these were the hallmarks of Woolleyian mystique. The sense of honor, the kowtowing, the flattery and hyperbole. I had flown there to perform the simple journalistic act of interviewing a man with a compelling story. But the endless cajoling made me wonder if I was somehow being played, the way Woolley once pitted reporters against one another to gin up coverage. When it came to Woolley's entrance, this too was melodramatic. Several weeks before my trip, Woolley, then eighty-three, had collapsed in his apartment and was rushed to a nearby hospital. He was supposed to stay longer, but when he heard I had arrived in London, he "jailbroke," he said, and convinced a nurse to summon an ambulance to drive him home.

I sat in Woolley's living room for more than an hour as we waited for Woolley's ambulance. I watched a clock on the wall that had yellow tufts of hair above the eleven and one, meant to resemble Woolley's symmetric tufts. Woolley's friend Gary, who first connected me to the man, sat across from me, along with a fellow named David, who did magic tricks and told dirty jokes. It was cold and rainy outside, and every few minutes Dave asked if I wanted my tea refreshed.

Between magic tricks, I looked around the apartment. Every inch of wall was covered by framed news articles, models of ships, certificates, and tchotchkes. Above the mantel was a framed photo of Woolley's former dog, a Welsh border collie named Flint, who died in 1997 and to whom Woolley dedicated his book. Every crevice of the apartment had been taken up by papers and memorabilia, much of it nautical and almost all *Titanic*-related. Before I arrived, I worried that Woolley might have thrown away old records that could be revealing. But it turned out to be the opposite. Woolley hadn't thrown anything away in decades, and now, as David kept darting to the kitchen every few minutes to make new tea, I had to tuck my legs each time he passed as he waddled between piles and papers to the kitchen.

Woolley moved into the small one-bedroom in 1982 in the neighborhood of Goodmayes in East London. The rent was cheap, and his friends lived nearby. Having once spent weeks sleeping on a park bench, he didn't mind that it was a working-class neighborhood with large pockets of South Asian immigrants who filled restaurants and pubs with food and music unfamiliar to him. He was too busy at the time watching Grimm's ill-fated expeditions and plotting his next move once Grimm found the ship. Woolley's continued appeals to raise the *Queen Elizabeth* and the *Titanic* didn't get much traction in the early eighties, particularly after Grimm began to siphon off his media. Because of this, Woolley spent those years working on the litigious side of the wrecks, writing letters, building partnerships, and threatening to sue anyone who infringed on his salvage rights.

Finally, Woolley's ambulance arrived, and he entered the room slowly, supported by a cane. He looked at me, let out a big laugh, and called me a "feisty bugger." I tucked my legs as he passed, and he shuffled to a chair positioned in the very center of the room—his captain's chair, he said, from which he controlled his operations. In front of him sat a pile of papers and, on top of it, a small laptop.

Now in his eighties, Woolley had retreated from the vigorous work of his forties and fifties. But it was easy to imagine Woolley as a younger man in the same chair in the same position, feeling like he was directing the world's traffic. Letters, phone calls, reporters, lawyers, threats, agreements, and promises. All keeping him in the mix and getting him nowhere at all.

"You know I have a personal tie to it," he said. "My two aunts, they were supposed to be on it but had a premonition."

He pointed to a framed photo of the ship on the wall above my head. He turned around and showed me scale replicas of more than four ships, including the *Queen Elizabeth* and another ship with the number *476*. His bookcase was filled with titles like *The Titanic for Dummies*, *Titanic: Triumph and Tragedy*, and *The Complete Idiot's Guide to the Titanic.*

He paused a moment and took a breath.

"Bloody hell it's grand, you being here," he said. "My mother once told me I'd never amount to anything, and now look at this, everyone wants to hear from me!"

This aw-shucks facade was also classic Woolley. He had nursed his underdog status his entire life, seeming to revel in low expectations that he could exceed with the most minor of accomplishments. No one believed he could actually own the *Titanic*, so when he declared he did, it was news. Anyone could dismiss the engineering ideas of a man with no engineering experience, so when he built a team, it was surprising. Each time he might have faded to obscurity, he, like the *Titanic* itself, received a burst of publicity that renewed the story and his place in it.

For several hours Woolley held forth on his long association with the *Titanic*. It had all started on a dare, he said, outside of a chip shop one night after he had talked for so long about the shipwreck that someone dared him to do something about it. He explained how he had gone about getting the salvage rights by putting the announcements in newspapers and waiting for objectors, who never came—a fact I later told him I couldn't verify and he didn't argue. He listed all the companies and legal entities he had created to manage the expeditions and salvage attempts.

"I was contacted by people all over the world!" Woolley said, his voice rising with both pride and disbelief in his own story. He weaved a long yarn about his visit to Hong Kong, the great trip of his life, though it had ended by almost any measure in disappointment.

Before meeting Woolley, I struggled to comprehend profound failure, of what it was like to look back on the dominant goal of one's life, a goal so specific and public it was printed in millions of newspapers around the world, and know that you had failed. I remembered hearing about a similar condition that erodes the spirit, when someone has a major success too early in life and lives the rest of their days looking downhill. This pathology was known to affect astronauts, particularly the three on Apollo 11 who walked on the moon as young men and returned to earth with their biggest accomplishment behind them and decades ahead to figure out what to do next.

But Woolley's pathology was different. His wasn't the quick success followed by an emptiness of purpose. His purpose was clearly defined, a linear arrow from boyhood. Yet it had never manifested into anything real. His failure had come in waves, a steady stream of setbacks that left him drenched in the whitewash of defeat. I pictured him like a lonely castaway, desperate to get to the open ocean yet unable to make it past the reef.

Strangely, though, sitting with me now was a man old and

infirm, but not one diminished in spirit. And as we kept talking, I started to get the sense that Woolley didn't believe he failed at all. Finally, I asked him bluntly:

"How do you explain the fact that we're sitting here in 2020 and the ship that you promised to raise fifty years ago is still underwater?"

He glared at me. I felt as though I'd overstepped, that my question was a gratuitous journalistic dunk on an old man. But his response surprised me and made me realize that, yes, not only was I being played, but that Woolley earnestly saw my visit—and this book—as a shot at a few extra yards down the never-ending field of his life.

"It's not that I didn't succeed," he said with a grin. "It's that I haven't succeeded *yet*."

I raised my eyebrows.

"And that's where you come in," he said.

° ° °

*The New York Times* reported the discovery of the *Titanic* on September 3, 1985, in a small story on the bottom of the front page, under the headline "Wreckage of *Titanic* Reported Discovered 12,000 Feet Down." Two days earlier, a team of American and French researchers towed an underwater-imaging craft over what was unmistakably one of the *Titanic*'s boilers. The search took a different approach from Grimm's earlier expeditions; instead of looking for the ship's hull, the researchers theorized that the *Titanic*'s high-speed sinking velocity combined with the pressure gradient had resulted in a destructive collision and implosion that left a sweeping debris field. When they found the boiler, it was a bread crumb that led to the hull itself.

The strategy was only as good as the technology behind it. In the two years that had passed since Grimm's failed 1983 expedition, the French and American team had the advantage of newer sonar and imaging crafts developed by the U.S. Navy for the Woods Hole Oceanographic Institute and the French marine institute IFREMER. One of the crafts was named *Argo* by Robert Ballard, the young navy oceanographer who had earlier critiqued Grimm's expeditions and now was leading one of his own. The *Argo*—named in homage to the mythical Greek ship that carried Jason on his hunt for the Golden Fleece—had dozens of cameras looking both down and forward, along with bright incandescent lights to illuminate the deep sea in a way it had never been lit before.

"We went smack-dab over a gorgeous boiler," Ballard said in an interview with a Canadian TV station over the radio of his ship, the *Knorr*, while it was still in the North Atlantic. "It was just bang, there it was."

According to Ballard, the crew erupted in cheers. Then the mood turned quiet once it set in that they had discovered the mass grave where fifteen hundred people had died. Someone suggested a moment of silence.

"We realized we were dancing on someone's grave, and we were embarrassed," Ballard said. "The mood, it was like someone took a wall switch and went click. And we became sober, calm, respectful."

In an image captured of the moment of discovery, Ballard is wearing a blue jumpsuit. He's standing in front of more than a dozen TV screens with his legs wide and arms crossed. His eyes are squinted, and his mouth hangs open slightly, the presumed combination of satisfaction and disbelief.

Years later, it would come out that Ballard and his team weren't quite as smart or as lucky as it seemed. What the public didn't know at the time was that Ballard had made a bargain with a senior Pentagon officer named Ronald Thunman, who offered to build the *Argo* for Ballard knowing it would help him find the *Titanic*. But in exchange, he wanted Ballard to first use the *Argo* to find two Cold War–era U.S. nuclear-powered submarines that had sunk in the North Atlantic. The two subs, the USS *Scorpion* and USS *Thresher*, were lost in 1968, together killing two hundred twenty-eight crew. The *Thresher* was profoundly tragic, having malfunctioned and sunk accidentally during deep-diving tests. But the *Scorpion*, whose sinking was attributed to an unintentional torpedo activation but has never been fully confirmed, was deemed more dangerous because it carried two nuclear weapons.

For more than a decade, American military leaders worried about the nuclear material on the *Scorpion*. But more concerning, they fixated on whether the Russians might have found and pillaged either sub. Ballard was a navy-commissioned oceanographer who had publicly discussed his interest in finding the *Titanic*, which gave him the perfect cover story to look for the submarines in the North Atlantic without tipping off the Russians. Under the agreement with the navy, Ballard would spend two months in the North Atlantic. Once he found and surveyed both subs, he could use any remaining time to search for the *Titanic*. In the end, this amounted to twelve days.

Like Grimm, Ballard had performed extensive modeling to deduce the precise wreck site. He accounted for "underwater wind," as he called it, to explain the distance lighter objects might have traveled as they took longer to sink, such as clothing from the cabins or spoons from the salon. A French team before him had come extremely close, missing the wreck site by fewer than one thousand feet, and making Ballard's place in history possible. If not for a gust of wind that pushed the French ship off course, the man forever credited with finding the *Titanic* and who would enjoy a lucrative career of speeches, books, and consultations with movie producers wouldn't be Ballard, but a French oceanographer named Jean-Louis Michel. (In the years after, Ballard occasionally referred to Michel as his "co-discoverer.")

The early coverage fixated on the photographs taken of the degraded ship. There were images of a propeller, a window from an officer's cabin, and a pile of more than fifty plates and bowls from the kitchen. A pot was visible in one shot and in another a stoking port for one of the boilers. Virtually all were grayscale images of corroded junk that would be easily dismissed if they didn't belong to the famous shipwreck. The photograph that attracted the most attention was taken in front of the bow, revealing the ship's recognizable visage, seen millions of times in books and movies, now appearing old and frail. The sturdy railings had worn down and were now covered in kelp-like debris and rust, looking like some underwater monster had sneezed on them. Appearance-wise, the ship had seen better days. But the photos confirmed the most crucial factor: it was the *Titanic.*

Ballard returned to Massachusetts in a way Grimm had coveted: in triumph. He also unleashed a new era of science. The discovery of the ship was manna for historians and shipwreck hunters, who in the following years would use a craft called *Jason* and several others to study every inch of the wreck, piecing together a detailed timeline of how it sank, why it imploded, and what happened to every item and person on board.

But among marine scientists, the *Titanic* seemed to be secondary. The bigger advance seemed to be the technology behind the *Argo* and the other crafts that had done the underwater survey and were poised to look farther and deeper than previously possible.

"This allows us to open up deep-sea exploration on a much, much larger scale than before," John Steele, the director of Woods Hole, told *Time* magazine. "We couldn't ask for more."

° ° °

The same way time is divided between major events—before Jesus and after, before Columbus and after—there was the era before the *Titanic* was rediscovered, and there was after.

Seeing the wreckage in photos sparked a new chapter in the forensic investigation of the disaster the same way a new burst of evidence can jump-start a criminal cold case. Almost immediately, proof that the hull had in fact broken in two offered a new lens through which to judge eyewitness testimony. Plenty of trustworthy people had claimed it hadn't. Were they lying? Was some phenomenon of trauma psychology at play?

Also evident was the depth of the PR effort undertaken by the White Star Line in the years after the disaster. A high-profile tragedy was bad enough for business, but the company had gone to considerable lengths to make the public record reflect that the *Titanic* sank intact and on an even keel. Any evidence that the ship broke up on the surface or imploded on impact would invite allegations of poor construction, which would mean a crushing number of lawsuits from survivors' families alleging neglect. In the wake of the disaster, the White Star Line was still operating the *Olympic*, built in the same design as the *Titanic*, and another, the *Britannic*, scheduled for launch in 1915. The more the public blamed the tragedy on an iceberg that no longer existed, the less the White Star Line or the ship's builders would have to answer for any mistakes.

Images from the 1985 expedition and a subsequent visit Ballard made a year later revealed how cynical the corporate effort had been, although after seventy-three years, it was also clear that it had worked. Evidence about the ship's faulty design, its weak midsection, and its suboptimal iron might have been bombshells in 1912, but in 1985, there was no one left to hold accountable, and anyone directly affected was dead or old. Instead, the overwhelming public reaction to the images in the eighties seemed to be disappointment. The optimistic hope for decades that the *Titanic* was preserved in a low-oxygen and cold-water environment was punctured by photos that showed that the ship looked awful. The grand staircase, the first-class cabins, the telegraph room—familiar sights from old photos—had been broken and scattered during its plunge and impact. Large growths of rust covered every metallic surface. Many of the decks had collapsed on one another. Every ounce of organic matter, including corpses, appeared to have been eaten by microbes.

"In a way I am sad we found her," Ballard wrote in the December 1986 edition of *National Geographic*. "Though still impressive in her dimensions, she is no longer a graceful lady . . . Her beauty has faded."

In public, Ballard compared the ship to an old girlfriend he corresponded with for years and with whom he was disappointed when he finally met her face-to-face. But in off-the-cuff moments, he was blunter about the reality that his legacy was now intertwined with something so ugly. "It's sort of like I just married someone," Ballard said to a reporter for *Newsweek*. "And is this something I want to be married to? It seemed nice at the time—you know, she was cute, she was nice and all that sort of thing—but now I'm married to her and wondering if I made a mistake. And I can't just walk away from this one. She won't let me." (I offered Ballard the opportunity to update this uncouth description of both a wreck and a fictional woman, but he refused. A friend of his told me that misogynistic language remains widespread among wreck hunting, which likely plays a part in excluding women from more prominent positions in the field.)

Nevertheless, Ballard's 1980s disappointment with the wreck's condition was undoubtedly short-lived once he realized the gold mine he had happened upon. Not only was he instantly the eminent expert on one of the twentieth century's biggest news events, earning him endless offers to write books, appear in films, and give speeches. It became clear—as Jack Grimm knew—that everything about him and his expedition could be sold. On September 10, 1985, less than two weeks after the discovery, Ballard confirmed plans to sell certain iterations of the underwater technology he used to find the *Titanic* to any willing company or government for as much as $15 million. The *Argo* and other crafts were developed and owned by the U.S. Navy, but the money was too good to stop Ballard from the most American of capitalist moves—taking a taxpayer-funded technology and repurposing it for private gain.

In his early forties, Ballard was an instant elder statesman of ocean exploration. But like any lottery winner, he struggled to reconcile his new life with his earlier values as an explorer and environmentalist. Selling sonar and imaging tech to help others make new ocean discoveries sounded nice, but the likely customers were oil and mining companies with little interest in archaeology or history. News of the discovery sparked interest in visiting the wreck to collect artifacts that could be auctioned. Some even debated anew raising the wreck using inflatable bags. Ballard suggested leaving the wreck alone in respect for the victims, and if not for them, at least for the sake of history. "You don't go to the Louvre and stick your finger on the *Mona Lisa*," Ballard said years later, still peeved at the commercial treatment of the wreck site. "You don't visit Gettysburg with a shovel. These guys are driven by greed." At the time, however, while Ballard was enjoying *his* payday from his association with the wreck, his plea for others to back off rang hollow.

If he wanted, Ballard could have ended his career right then and spent the rest of his life telling the same discovery story on the never-ending circuit of rubber chicken dinners. But there was an extra benefit to finding the *Titanic* he hadn't expected.

Before the discovery, he had spent years studying hydrothermal vents and the physics of the seafloor. He had made discoveries as elemental to science as the possible foundation of human life. It wasn't until the *Titanic*, however, that the letters started arriving on his desk that added new motivation to his hunger to keep looking forward, to use his privileged position and celebrity to go find other lost relics of human exploration, a quest that would lead him to the German battleship *Bismarck*, the American aircraft carrier the USS *Yorktown*, and the second-most-famous passenger steamer in history, the *Lusitania*. He would start his own institute for ocean exploration, launch his own research boat, and, even as of the writing of this book, be hunting for the mysterious remains of Amelia Earhart.

In those days and still today, the letters were almost all about the *Titanic*. Almost all were from kids, and they all said the same thing:

"What do I have to do to do what you do?"

∘ ∘ ∘

Doug Woolley can't remember where he was when he first heard about Ballard. But sitting in front of me, he grimaces every time I mention Ballard's name. He tells me that Jack Grimm had been courteous and respectful enough to seek Woolley's blessing and partnership before his expeditions, but Ballard hadn't bothered to call the man who had been quoted in thousands of newspapers as the self-proclaimed owner of the shipwreck.

Before meeting Woolley for the first time, I hadn't had a chance to consult the lawyers and salvage experts about whether there was any truth to Woolley's ownership claims, or whether a person even could own a cultural piece of mangled steel. I took Woolley mostly at his word—not that the *Titanic* was really his, but that he believed it was. I asked him for documents, court judgments, even copies of the newspapers where he planted his stake and gave everyone two weeks to object.

"It's all in there," he told me, pointing to a set of scrapbooks on the shelf.

I spent an afternoon flipping through the albums, each one stuffed with hundreds of news clippings dated as far back as 1967. I started taking pictures of the articles to read in detail later, but after the first fifty or so, I stopped. There were too many.

The earlier sense I had that Woolley was an eccentric shipwreck obsessive was belied by the fact that for decades people took him extremely seriously. There were articles in Spanish, Chinese, French, and German. Kids wrote him letters from all over the world, and he was invited to talk at universities across England. Woolley never got his salvage plans even remotely off the ground, nor did he ever truly own the *Titanic* in the way someone owns a car or even a pair of shoes. But it was surprising how little that seemed to matter. Woolley believed it, the media hyped it, and the people who read newspapers in dozens of countries bought it. A big lie can bring big consequences. But in his case, the fact that the ship was entirely inaccessible for so long reduced a costly lie to a mere falsehood of no consequence.

This changed, however, when word reached Woolley that Ballard's team had found the ship. For one, the discovery burst the bubble for Woolley himself, who realized he was completely passed over in the tidal wave of Ballard's news coverage. Woolley wasn't mentioned at all, nor was he consulted, credited, or quoted in the following years when additional expeditions collected thousands of accessible artifacts for museums and private collectors. One wonders if the realization in Woolley's mind came slowly or all at once that he didn't own the *Titanic* and never had.

The mind has a way of performing gymnastics to placate the heart. And before I could ask Woolley this pointed question about when it all crashed down, the tone of our conversation shifted.

"The missing link is *you*!" he bellowed with a big belly laugh. As he cackled, he pointed at me and looked around the room. "That's why we were so happy when you called to visit."

Everyone in the room was laughing now, but I didn't follow the joke. Woolley leaned over in his chair and put his hand on my shoulder.

"Don't you see?" he said. "We need your help."

I still didn't understand. But in the next half hour Woolley transformed from answering my questions to forcefully lobbying me to join his team—to, as he put it, "get the job done."

Sometime later, it finally clicked that the 1985 discovery of the *Titanic* resulted not in a sad and private deflation for Woolley but in another moving of the goalposts. When news reached Woolley of Ballard's success, Woolley first denied it was true. When he learned it was, he seized on one peculiar detail—that the "moment of discovery" had occurred in the middle of the night, when Ballard was asleep in his bunk. Woolley argued that ergo, Ballard's discovery story was based on a lie. (Ballard later elaborated that yes, he was in his bunk at midnight, but he couldn't sleep, so he wandered back to the bridge before the wreck was confirmed.) When Ballard's leadership role was cemented in the press, Woolley evolved to embrace the discovery as legitimate but claimed victory for *himself*, declaring that Ballard "did my work for me." The final transformation of Woolley's rationale was that it was actually good that someone finally found it—and even better that it wasn't him—because now the path was clear for him to accomplish his bigger goal: to salvage the wreck and gift it back to the world.

In this context, the autobiography Woolley published in 1998 made sense not as a memoir of his failed exploits but as a blueprint for what he saw as still possible. The book was less a vanity project than a marketing tool to rally support and money for the lifting operation, and he rushed its production to catch the wave of *Titanic* enthusiasm after James Cameron's 1997 film reignited the story of the aging wreck. The book, however, sold poorly, mostly on account of Woolley's anemic marketing. But another bolt of luck came at the perfect time. The arrival of the internet and the birth of Amazon kept the book in perpetual circulation and Woolley's association with the wreck perennially searchable. Starting in the year 2000, every few months a *Titanic* nut would see Woolley cited on some website and call him up. Once, a young man called and invited Woolley to do an interview, which the man then posted on YouTube, where it still lives. Another time he was invited to a club in London for shipwreck lovers and was treated like a celebrity. By holding on to the wreck long enough, he secured himself a place in the long and never-ending story of the *Titanic*.

This wave crested in 2013, when someone from the TV show *Salvage Code Red*, which aired worldwide on the National Geographic Channel, contacted Woolley in hopes of featuring him on an episode of the show that described the technicalities of raising a century-old wreck. Woolley was giddy for weeks leading up to the filming and spent several days talking with producers. But ultimately, the episode was scrapped and never aired. Woolley nursed the sting of disappointment about the show for years, until 2019, when the next person to call him was me.

"I've sacrificed friends because of what I've done," Woolley said, a little wistful. "I sacrificed principles sometimes. It's not something you want to do. But if you believe in something, you do what it takes."

As I sat before him in 2020, Woolley believed the *Titanic* was still in salvageable condition and that the ship could be raised. He believed the steamer trunks of his two great-aunts, the ones scheduled to board the ship but spared by their foresight, were still underwater waiting for him to claim them. At eighty-three, he seemed to believe that everything he imagined as a young man in 1966 was still true, that all it would take was a little more time, a little more publicity, and a little more money to *finally* get the ship out of the water. Woolley even calculated the precise amount he'd need—£18 million, or about $25 million—and he said he already found a salvage company in London that was standing by to do the job.

"Call them, they'll tell you!" he dared me. I eventually did, and the owner told me that Woolley had been in touch several times and he simply didn't understand the physics of deep-sea engineering.

I walked back to my hotel that night in a fog, turning over in my head how a man's lifelong dream to do something almost certainly impossible was now my responsibility. But mostly I felt sad. Here was a man who committed his entire life to a pipe dream, an insane and outlandish plot that he convinced his close circle of friends to believe in as much as he did. How many people had fed his delusion?

And how many people had laughed in his face and told him the whole idea was ridiculous?

I left London deflated. I had failed to get a story from Woolley rooted in reality, while at the same time I had unintentionally raised the hopes of a vulnerable old man who believed that I—and the publication of this book—would be the final domino that would set in motion the climax of his life's work. I didn't have the heart to tell him it wouldn't, nor did I have the courage when we spoke on the phone weeks later to admit that I hadn't asked "the shipwreck department" at *National Geographic* to call him back. (There is no shipwreck department.)

But just as it always does, the tide began to change. And the more I talked to Woolley on the phone and read his old press clippings, the more I started to see something different. Here was a man born to a poor and broken family, mocked by his own mother and taught as a boy the cruelty and loneliness of the world. He worked in drudgery for decades in low-wage jobs, monitoring train tracks or making women's pantyhose. And all the while, instead of growing bitter, he weaved his own security blanket. He declared a north star of his own that he could always turn to and run toward, even if he never fully reached it. As he grew older, the security blanket from childhood had become a full-size parka of emotional armor. The *Titanic* brought not just a source of credibility and fame to a man accustomed to neither. It was also a reliable receptacle of energy. There were always new details to find, memorabilia to collect. He had a reason to call strangers in other countries and forge friendships. The *Titanic* was his calling card, a story both eminently famous and limitless to explore.

It's little wonder why Woolley ended up at the center of the tale, at least for a while. The sheer repetition of his association with the ship bore fruit, and once it did, could he really stop? Could he really be expected to call a press conference and declare to the world that things had gone too far and he had no business talking about advanced matters of law, engineering, and nautical history? I had initially pegged Woolley as a relentless hype man running a long con too good to give up. But what if it was the other way around, that Woolley was hyping himself? That over time, he had shoveled so much coal into the boiler of his obsession that to walk away would risk the whole thing exploding into the sober realization that it was all a big waste.

There's an old Hemingway quote that "a man is never lost at sea." The first time I heard it, I dismissed it as a poetic lyric that ignored the millions of people who have indeed been lost at sea. But I realized it could be a new lens through which to see Woolley. Through his life's many chapters of war—poverty, homelessness, solitude, heartache, embarrassment, and endless mockery—he never seemed to be alone. He always had somewhere to be and someone to call to keep the project moving. He always knew what he would be working on the next day, the next month, and the year after that. Does it matter if a man's path turns out to be a treadmill so long as he puts in the effort to move himself forward?

One can't call Woolley a success, particularly after falling so short of his publicly declared goal. But neither does a failure seem right. Life is rarely so binary, and anything related to the ocean comes with extra degrees of nuance.

Woolley probably occupies a rung somewhere in the middle, his story benefiting from the fact that he hasn't declared it over yet—and never will.

You can label his life as a case study in naiveté, fantasy, fiction, and delusion.

Or you could simply call him a dreamer, the kind we need more of.

Chapter 13

# A REDDISH STAIN IN THE MUD

For a century after the *Titanic* sank, Harland & Wolff, the shipbuilding company that made it, went silent. What began as an effort to avoid legal liability for an unsinkable ship that clearly wasn't morphed into a corporate shame that as the years went on nearly shrouded Northern Ireland in cultural embarrassment. Harland & Wolff continued building ships after the *Titanic*, first gunships and cruisers for the First World War and then aircraft carriers for the Royal Navy to fight Hitler. It built a reputation for repairing damaged ships in the fifties and sixties, but it never quite shed its black eye from the *Titanic*. For a hundred years, anyone who called looking for an interview, photographs, or a tour of the dry dock where the ship was built rarely got a call back.

That is, until one day in 2010, when an emissary for the British royal family quietly contacted Bob Ballard. James Hamilton, the

Duke of Abercorn, said he was ready to talk and that Northern Ireland was ready to declare the *Titanic* a distinctly Irish tale.

"It's been a hundred years that we've basically denied we built the *Titanic*," Hamilton told Ballard. "We now want to embrace it and tell the story."

He wanted help from *Titanic* experts like Ballard to build a credible museum about the disaster. Over four years, the government quietly outfitted the *Titanic*'s former dry dock with railings, walkways, and exhibit boards. Officials enlisted the famous British architect Eric Kuhne to design a visitors' center so distinct it would rival other iconic structures like the Sydney Opera House and the Guggenheim. Kuhne's idea was a structure with four wings protruding from a center point, each like bowed hulls of a ship and covered with thousands of multicolored aluminum shards to give the appearance of the sea. Some early visitors nicknamed it "the iceberg."

There's not much to see at this museum, at least for anyone hoping to see the actual ship. Artifacts sometimes pass through on their never-ending journey between featured exhibits. There are several permanent installations of wood and metal excavated from the wreck. The closest anyone gets to the recognizable *Titanic* is through video galleries that allow kids to feel as though they're standing in the dining room or the rebuilt grand staircase, where people take selfies. A cartoon mascot of Captain Edward Smith wanders around, and everyone poses for photos at a rudderless helm or below signs in the cafeteria cautioning, WARNING, ICE AHEAD.

Yet still, the experience tugs somewhere deep in the human hunger to be close to something famous, even if the famous thing never shows up, like visiting Graceland to be near Elvis. Belfast officials thought the museum would attract half a million people a year. In its first year, it almost doubled that. Wanting to drum up money and visitors out of nowhere, the museum followed the well-worn track of exploiting the *Titanic*—and the ship, as it always does, delivered.

For almost all other ships, measuring a wreck's value means adding up the price of its contents, a record long held by the *Nuestra Señora de Atocha*, the 1622 Spanish galleon that sank with up to $400 million. Nothing can compete with that. But measuring the full reach of the *Titanic* economy is like trying to estimate the all-time value of the color blue. In addition to the well-known movies, books, and documentaries—collectively worth billions—there are entire universes of fan fiction, performative interpretations, video games, and an endless quantity of collectibles created out of thin air. At any given moment, eBay hosts more than a thousand auctions for questionably authentic items from the *Titanic*, including wood allegedly from the ship, replicas of china, posters, certificates, coins, tickets, and, most curious of all, a nickel-size rusticle from the wreck site sealed in a plastic case (starting bid: $995).

One day I came across the *Titanic* Store, a website that sells replicas of more than one hundred fifty *Titanic* artifacts. I called the customer service number and talked to an energetic man named Ozgur, who insisted that his site was the "real" *Titanic* collectibles store. "Anybody can make a plate or teacup from the boat, but we try to respect the story," he said. His top seller was a tiny piece of coal apparently collected from the *Titanic*'s boilers in the nineties, each with a certificate of authenticity. His online store is owned by Experiential Media Group, a live-events company in Atlanta that, after a long string of mergers and takeovers, now owns and manages the largest collection of authentic *Titanic* artifacts, estimated at nearly five thousand items. The collection began with the earliest dives in 1986 and 1987, which, contrary to Ballard's pleas, pulled up thousands of pieces of detritus from the wreck and the debris field. Experiential puts them on display year-round and occasionally in the hands of school groups, fueling the next generation of *Titanic* obsession.

Between the Belfast museum and the artifact collection in Georgia, you can see the outlines of the educational grave site Doug Woolley dreamed of. Rescuing the wreck and towing it back to dry land would have earned him fame, but his bigger goal was to put the ship on a literal pedestal to honor its victims. His vision morphed over the decades, but what stayed constant was the intention not just to surface the ship but to set it down somewhere people could see it and touch it, a permanent vigil for a long-faded tragedy.

° ° °

Today, depending on where you look, there are different *Titanics* at the wreck site. The recognizable bow shows accelerating corrosion from salt water and rusting. New holes in the metal can sometimes be seen from one visit to the next. The holes not only weaken the broader structure, they also expose inner crevices to relentless water currents that have been known to pull apart remaining portions of the ship. Except for some carvings made of hard and erosion-resistant teak, the rest of the woody organic material was fully chewed up by bacteria by the end of World War II.

Other parts are in much better shape. After the bow's fall, it struck the seabed at an angle between fifteen and thirty degrees, driving its pointed end about sixty feet into the mud. Buried for a century, this portion is thought to be pristine, with its original coat of paint. But it will never come out. Not only is the exposed steel on the upper bow too brittle for even the most industrious crane operation, but the mud has also acted as deep-sea quicksand for longer than most humans have been alive. Pulling it up would require the force of an interplanetary rocket. Even if it was possible, it'd be counterproductive. Exposing century-old steel to the elements would kick-start its instant breakdown.

There are still several decks, hallways, and staterooms in the interior of the ship. Researchers assume there could be existing furniture, old mattresses, and passenger mementos inside the hull. Scientists sometimes debate whether there could be human remains buried deep in the hull and protected from the elements. But it's almost impossible to find out. The narrow structure is too cramped and brittle for autonomous underwater vehicles and, thus, human eyes.

The stern, meanwhile, also lodged itself in mud, about fifty feet deep. But the rest has been leveled flat. Ten decks have collapsed into a heap of loosely structured rubble less than twenty feet tall. Any hopes of seeing the innards of the sunken stern were always a long shot, even the night it sank. As the decks collapsed on one another over the years, every crevice has been ripped open, exposed, and attacked. The two main propellers were almost yanked off the stern during impact and were bent upward. The smaller center propeller is completely buried in mud and has spent the last one hundred ten winters, falls, summers, and springs unseen and untouched, ensuring an endless debate over whether it had three blades or four.

The difference in condition between the bow and stern was a mystery in the eighties and nineties, considering both sank on the same night from the same spot. The discrepancy has since been chalked up to hydrodynamics. Both pieces fell at the same terminal velocity of about thirty-five miles per hour. But the bow's pointed end absorbed more trauma than the stern. The stern was also badly damaged from the start. The breaking apart on the surface pulverized its underbelly and pulled out several of the boilers, the heavy masses that gave the ship its ballast. Without the boilers, the stern had the structural integrity of a house without a foundation.

The most evocative and recognizable part of the *Titanic*, the grand staircase, no longer exists. Built from oak and installed step by step, it is believed to have been ripped from the stern as the ship sank and broke apart from the stress. Another theory ponders whether the staircase made it to the seafloor only to be obliterated by the down-blast of water, like a daffodil under a firehose. In a book about the making of the film *Titanic*, James Cameron revealed that while shooting the flooding scenes, the replica of the wooden staircase dislodged from the floor and unexpectedly floated up. He suspected the same thing happened to the original.

Outside of the two main pieces, the remaining forensic clues about how the ship sank lie in the scattered mess of clothing, dishes, suitcases, toothbrushes, boots, and broken glass. The *Titanic*'s debris field is actually five different fields, each linked to a different moment in the sinking. The north debris field is the result of the bow's impact. The central field is from the stern's. The remaining three are items that dislodged on the surface or fell off on the way down.

Shipwreck investigators refer to this as a "crime scene," even though there's no real crime. Compared to thousands of other ships that have sunk with no trace, there's no debate over where and when the *Titanic* went down, at least not anymore. Instead, the granular study of an aging wreck is a work of reverse engineering—plugging items into a model and then rewinding the tape. This method helped solve the long-standing question about how the forward hatch cover that was bolted down ended up more than one hundred fifty feet away from the hull.

While there's only one *Titanic* wreck, there are many ways to interact with it. Several years ago, a German artist named Yadegar

Asisi offered an artistic interpretation inside a defunct natural gas silo in the city of Leipzig. Asisi projected a life-size image of the *Titanic* wreck as he *believed* it looked in a circular image that spanned thirty-seven thousand square feet and was bathed in hazy blue light. He placed suitcases, mangled furniture, and old ropes for rigging. He distressed the upper cabins to look like they had gone through a hurricane, and he asked a musician to compose a moody soundtrack that featured "background noises tailored specifically to the mysterious mood." For several weeks, visitors stood on a raised platform gawking at the airbrushed and fictional depiction of the wreck.

Another option—one that leaves nothing to the imagination—is to see the wreck site in extreme, almost overwhelming detail. Since the location became known in 1985, virtually all visitors to the site have been documentary filmmakers. But in 2010, just shy of the *Titanic*'s centennial underwater, scientists with Woods Hole, NOAA, and the National Park Service decided to survey the entire wreck site and convert the data into a 3-D map.

This amounted to a dramatically different way to look at a wreck. While filmmakers focused on the most recognizable parts, like the bow and the stern and their contents, the Woods Hole scientists wanted to treat every inch of the wreck site as equal. They deployed a new vehicle that ran acoustic transects back and forth over the sprawling site, a process known as "mowing the lawn." A different vehicle collected optics, hovering for hours at a time to take thousands of photos and high-resolution video.

They expected that when the data was compiled they'd see the kind of 3-D image you get when you go to the movies and have to wear silly colored glasses. But the data revealed a view far better than looking through the porthole of a submarine. You could zoom in on any part of the wreck, flip it around, and inspect the same spot from another angle. The data from the 2010 Woods Hole visit and several more that followed was a forensic gold mine. The biggest immediate takeaway was that the bow and stern were significantly more damaged than decades' worth of 2-D photos let on. Seeing the damage up close helped investigators reverse engineer the ship's fall through the water column and model its impact with the seafloor, the destructive downblasts of water, and the explosion of debris. What was more, comparing the 2010 imaging with data from future missions would provide a temporal fourth dimension to watch the breakdown and forecast its future pace.

The hull appeared to have sagged since the last time anyone visited. The rusticles were fragile, prone to turn to dust at the softest disturbance. The rusticles also extended into the steel structure of the ship, like roots of a tree. In 2010, Henrietta Mann, a biologist at Halifax's Dalhousie University, discovered that the bacteria eating the *Titanic* and responsible for the rusticles appeared to be a unique species observed nowhere else on earth. She named them *Halomonas titanicae*. She also observed that the length of the rusticles corresponded to the pace of bacterial activity, and the rusticles were big.

Digitized images could eliminate the water, the darkness, and the floating dust, offering the full monty of a wreck you couldn't even see in real life. A group of filmmakers animated the data, placed it behind a dramatic soundtrack, and spliced in expert interviews. Both Discovery and the National Geographic Channel aired the finished documentary, called *Drain the Titanic*, which repeated, over and over, just how remarkable it was to see the wreck up close.

That said, even the best graphics can't compete with seeing the real thing with your own eyes. Private tour companies spent years after the 1985 discovery taking visitors to the patch of ocean above the wreck to sprinkle rose petals on the water and bow their heads in silence. In 2001, a New York couple gave new meaning to their destination wedding and got married in a submersible hovering above the bow. Tourism declined in the early 2000s on account of limited and cost-prohibitive submarines available. But in 2021, a company called Ocean Gate Expeditions began selling tickets for midsummer "citizen science expeditions" for $150,000 a person. Each traveler would be given the title of "mission specialist" and the chance to listen to lectures by *Titanic* experts, and, weather permitting, take a trip to the wreck site.

Tourism is good for the *Titanic* brand, but it's also its undoing. "Visitors do more damage than anybody else," Ed Kamuda, the president of the Titanic Historical Society, said in 2003 in a statement that could describe any tourist site on earth. Fame brings exploitation and prioritizes volume, perhaps best illustrated in 2019, when a sub accidentally collided with the wreck. Today, strewn along with the old debris, are a growing number of beer bottles, miles of fishing line, and piles of old nets. There are weighted chains dropped by autonomous underwater vehicles, along with dozens of plaques, bouquets of plastic flowers, and padlocks tossed overboard by lovers. Eventually there will be more pieces of trash around the *Titanic* than pieces of the *Titanic* itself, if there aren't already.

◦ ◦ ◦

One of the last great debates about the *Titanic* is how much time it has left. When Ballard first saw the wreck in 1985, he was apparently "amazed," all things considered, at how well-preserved it was and declared that, if left unmolested, it would change little in his lifetime.

In 1995, Henrietta Mann, the longtime *Titanic* researcher, compared photos from the prior decade and gave it thirty years.

Five years after that, a submarine pilot named Anatoly Sagalevich, who hovered around the wreck for hundreds of hours with the filmmaker James Cameron, thought the wreck had two years left, if that. "This is terrible," Sagalevich said. "I have never seen the *Titanic* in this bad condition."

And yet, just like the *Titanic*'s story refuses to fade, the wreck continues to buck expectations of its demise. When the wreck was still chugging along in 2010, Woods Hole scientists said it was in surprisingly decent shape and proposed a new visit to the site to chart its breakdown.

That same year, 2010, was also when Mann and her colleagues discovered the *Halomonas titanicae* bacteria that was accelerating the iron decay. Mann said the breakdown would happen "much faster." In 2019, she gave it, again, thirty more years.

The scattershot predictions reflect the complicated reality of every wreck. It's tempting to think of a ship as a single cohesive structure breaking down at a steady pace, like an ice cube on hot pavement. But disparate materials behave differently. Steel breaks down slower than wood. Glass and cotton, porcelain and brass, apples and oranges. No one can doubt the ship is wasting away, but declaring its dying day is a question not of the whole but of its parts.

It also depends on the definition of *gone*. As a working ship, the *Titanic* is already long departed. As a wreck, it's on a fast path to blurred disintegration. The most senseless irony may be that the longest stretch of the ship's life-span will be neither as a ship nor as a wreck, but as a pile of rubble. Like a house flattened by an earthquake, it'll take centuries before all that's left is an empty field of dirt.

And when will the last day come? When the last rusticle and the last leather boot—the tannins in leather having resisted most bacteria—flow into the current and all that's left is a reddish stain in the mud? The only sure bet is that none of us will be around to see it.

"It'll be there for hundreds of years," said the wreck hunter David Mearns, who has seen plenty of wrecks—some thousands of years old—that aren't going anywhere. The bottles, the leather, the metal, will be slowly eroded away. Structurally it will collapse. But there will always be a wreck site.

"People like to prematurely declare its death. But it'll always be there."

° ° °

In February 2013, a famous Australian man named Clive Palmer flew from Queensland to New York City to unveil a new grand idea. Palmer packed a tuxedo for a gala he planned to host aboard the USS *Intrepid*, a World War II– and Cold War–era aircraft carrier that in 1982 became the centerpiece of the Intrepid Sea, Air, and Space Museum on the west side of Manhattan. The museum houses, among other symbols of engineering triumph, a retired Concorde supersonic jet and the space shuttle *Enterprise.*

Flying halfway around the world was a business decision for Palmer, who happened to be a billionaire. The prior year, Palmer announced to Australian media that he planned to bring the *Titanic* back. Not by raising the defunct wreck, but by building it again.

Anyone who knew Palmer wasn't surprised by his ambition to build a fifty-six-thousand-ton, eight-hundred-eighty-foot-long, near-perfect replica of a hundred-year-old ship. Palmer had made his fortune in Australia by buying up land he believed held mineral deposits of coal, nickel, and iron ore, and when he tired of that, he turned to vanity projects and, as he called them, "big schemes." He bought golf courses and a soccer club and tried to start a new political party. He decided to build a dinosaur-themed amusement park next to one of his golf courses and commissioned one hundred sixty life-size models of dinosaurs. He placed the biggest, a thirteen-foot-tall T. rex, on the ninth green as an attention-grabbing gimmick. ("He's our salesman!" Palmer said.) Like any big dreamer, Palmer also had big flops. His dinosaur amusement park failed so spectacularly that it killed the entire town around it. His political party, the United First Party, faced deregistration in 2020 after it struggled to sign even five hundred members. His short stint in government ended in—according to one journalist during a brutal interview—"at best an embarrassment and at worst a disaster."

But the *Titanic* seemed to Palmer like a sure bet. People loved the *Titanic*; people would always love the *Titanic*. It was a cash cow for people for generations. Why not him?

In some ways, building a ship would be easier than building a theme park. Palmer calculated that, since it was a ship, it could be built anywhere, and if it could be built in China, it would be cheaper by a magnitude of almost ten. The price tag wasn't bad, either. No one knew the final cost, but estimates pointed between 200 and 500 million American dollars, which on the lower end happened to be the exact same cost adjusted for inflation that the original *Titanic* cost in 1911. Palmer offered to guarantee much of the money himself. But he needed financing, and that required a sense of assurance that a man who'd never built a ship knew what he was doing. Which brought him to a fundraiser in New York.

Palmer was a charmer, a man who could talk anyone into or out of anything. But the harder part of building the *Titanic II*, as he called it, was to actually build it—to make sure the ship stayed on schedule and on budget, that it mimicked every detail possible of the original ship, and that, when completed, it wasn't just Clive Palmer's cheesy vanity boat, but an exciting and respected addition to the cruise industry that would attract high-dollar bookings for decades. This was a complicated sales pitch that amounted to branding a relic of the past as an embodiment of the future.

"When the day comes, and it most surely will, when *Titanic II* sails into New York, you'll be able to say that you were here, you were here tonight when it all began!" Palmer bellowed at the gala on the *Intrepid*. After Palmer's speech, the crowd broke into a conga line. People waited to get their photo in front of a life-size photo of the *Titanic*'s bow.

Actually creating the *Titanic II*, however, would face instant red tape. Global regulations of cruise ships had transformed since 1912, many specifically because of the *Titanic*. The woody interior of the ship was a fire hazard. Rivets had been replaced by welded hulls that affected the shape of the bow. Even if there were enough lifeboats, they had to be closer to the water, which would make the ship's famous black abdomen look like it was wearing a skirt. A South African businessman named Sarel Gous had dreamed of reconstructing the *Titanic* a year after the ship's 1997 Hollywood rebirth. But jumping through so many hoops had proven a major buzzkill. With each concession to safety, efficiency, and liability, his supporters and would-be customers grew bored and moved on.

Palmer looked elsewhere for inspiration. If built, his *Titanic* would be the fourth *Titanic* replica to exist. The first two were built in the landlocked towns of Branson, Missouri, and Pigeon Forge, Tennessee, as tourism magnets. Both were constructed as *half*-scale replicas and only the front half of the ship. As of the publication of this book, the third *Titanic* replica is still under construction on the banks of the Qijiang River in Sichuan Province, China, where it will be the centerpiece of a new seven-star resort and amusement park known as the Romandisea Seven Star International Cultural Tourism Resort. The rural area around it has little tourism, but the *Titanic*—ever the reliable draw—is designed for international appeal. Still, its most frequent customers are likely to be Chinese tourists, who have been uniquely enchanted by the *Titanic* story since its Hollywood treatment. In 2012, after James Cameron's film was released in 3-D, the box-office haul in China was four times bigger than in America. The *Romandisea Titanic*, as it will be known, will be permanently moored. It will never move and never sail, which makes it more of a boat-shaped hotel than a working ship. There will be no boilers or navigation equipment, no telegraph, no crew, and no working lifeboats. It will also never sink.

Sinking is synonymous with the *Titanic*, and Palmer thought his ship needed at least the *potential* to sink. That would make his authentic and also more dangerous, which he imagined would be its biggest draw. In 2013, Palmer told an interviewer that seeing the *Titanic* was easy, but *sailing* on the *Titanic* and feeling the sense of opulence, merriment, and risk of ocean peril would be the stuff of bucket lists.

Despite Palmer's promise, it's hard to imagine the *Titanic II* ever rolling off a dry dock into the sea. In 2013, Palmer vowed that the ship would be done by 2016. In 2016, he said he needed two more years. Two years later, construction had yet to start. The project seemed stuck in a familiar loop of financial difficulties and legal squabbles. Investors were attracted to a project that seemed exciting before realizing it was a pipe dream built on fuzzy math and unrealistic expectations. After its hypothetical launch, the *Titanic II* would be instantly dwarfed by modern cruise ships with flashy un-Victorian amenities like water slides and skydiving simulators. A few people might like to dress up and pretend they're living in 1912, but would thousands of people? And if they did, on a ship with limited electricity, no TVs, and certainly no Wi-Fi, what would today's smartphone-addicted travelers do on days two, three, and four, after the novelty wore off and seasickness set in? Worst of all, the small number of cabins relative to today's ships would make ticket prices ten times higher than the average cruise vacation. Was a *Titanic* knock-off worth it?

Palmer is often labeled the Australian Donald Trump, a bravado-fueled businessman who has gotten further than his critics admit and who has failed more than his supporters believe. His record in business and politics is long enough to choose your own narrative. He is also, oddly, married to a much younger woman from Eastern Europe.

But Palmer also embodies a certain desperado I had come to know. A man who perennially longed to be taken seriously and, after a lifetime of overpromising and underdelivering, never quite was.

Like Charles Smith, Palmer was a master of earthly engineering who wanted to leverage his experience in mining into an even grander idea. And he had early success, to the point that strangers wrote him checks and lined up to be his first customers.

Like Jack Grimm, Palmer was a man of almost limitless means. He learned early in life the power of a gimmick and the truth in P. T. Barnum's maxim that the world is full of suckers. Palmer, like Grimm, wasn't looking to *pay* to bring back the *Titanic.* His money was an investment that the *Titanic* would eventually pay *him*. He was a prospector first and last, and the *Titanic* was a well that never ran dry.

Most of all, Palmer resembled Doug Woolley. For one, he was extraordinarily difficult to reach. I spent weeks working every angle to find Palmer. I talked to reporters who covered him, an investor who sued him; I even asked a real estate agent who once sold Palmer's daughter a house. A friend finally told me that Palmer was defeated and tired. After another embarrassing episode when Palmer bought millions of doses of a Covid-fighting drug that turned out to be mostly worthless, he began to fade back. A man hungry for endless attention seemed finally to have had enough.

But the *Titanic II* was still on, the friend told me. And sure enough, when I emailed the managing director of the project, I received a reply that the team was "overwhelmed with the support and encouragement from fans and supporters around the world."

This was also classic Woolley. A sense that, despite all odds and all evidence, the great promise hadn't come undone. It was still coming together.

Long gone is the grand ship that once sailed and sunk. The wreck will disappear, too. The final iteration will be as a character in the mind, a name notable only because it's been repeated as a relic of something that once meant something but doesn't anymore. After the full story of the ship is written, the oddest part will turn out to be not that it sank, but that sinking actually lengthened its life.

The *Titanic* lengthened Doug Woolley's life, too. "The only reason I'm still alive is because of the Big T," he told me. At every point he can remember, his life had purpose because of the ship. And in the off years, when momentum would flag or another person stole the limelight, something would always happen or someone would show up and spark new energy for him to forge ahead.

It was self-fulfilling, of course, and a bit sentimental. But if it worked, who can judge? Many have, and many more will, but Woolley got what he came for: the adventure of a lifetime, with him at the helm.

# Acknowledgments

In the process of writing this book, I came across the astounding statistic that somewhere around 90 percent of every item we encounter in our daily lives came to us, at one point or another, on a ship. As someone who can't handle even a canoe ride without getting seasick, my first thanks goes to people who crisscross oceans to make modern life possible.

My intrepid agent, Lauren Sharp, helped spark the idea for this book. And my eagle-eyed editor, Brent Howard, fueled it with enthusiasm, thoughtful questions, and good advice, including that I finish it *before* my son arrived. "It'll be harder after," he said. (He was right.) Evelyn Duffy was the editorial find of a lifetime who arrived at the perfect moment to laugh at (and then delete) my nautical puns, Andrea Monagle deftly helped fine-tune the final drafts, and Grace Layer at Dutton steered this top-heavy dinner cruiser through the narrow canal of production. Matt Twombly is the talented creator of this book's maps and diagrams. And unless you're friends with my mom or my mother-in-law, you're probably reading this because Jamie Knapp is so ridiculously good at marketing books.

I picked up loads of information from seasoned maritime and shipping experts. Bill Thiesen with the U.S. Coast Guard walked me through ship design and engineering, and Joe Mazraani and Jennifer Sellitti took time away from arresting shipwrecks to explain what it means to arrest shipwrecks. Parks Stephenson was, again and again, my ambassador to the *Titanic* underworld.

I met many people who spend their lives protecting our oceans. Bob Schwemmer is my local marine archaeologist in California, and he's also one of the best in the world. Chris Goldblatt with the Fish Reef Project helped me navigate through artificial reefs, as did Jonna Engel with the California Coastal Commission and Jordan Byrum with the North Carolina Department of Environmental Quality. Jordan gave me the excellent advice to call Tim Mullane, the wreck scuttler, who, like all of these people, rarely gets recognized unless he screws up.

Ian MacLeod could talk for hours about ship corrosion, and we often did. Linda Elkins-Tanton was confused but good-natured about why a book about shipwrecks needed an interplanetary geologist. Alana Fossa was the cheerful voice that answered the day I called the corrosion hotline (which, if you have a question about galvanization, is 720-554-0900). Charlie Carmona at Guild Labs was my guide to the value of buried treasure. Chris and Allison Selman fueled one of my favorite stories in this book, about their herculean efforts to halt the decay of historic wrecks. And that's to say nothing of the lawyers. I actually had the harebrained idea while writing the sections about salvage law that I'd acquire an old wreck of my own, a bit like Doug Woolley, that I could use to impress people at cocktail parties—until I talked to Scott Bluestein, Chris Johnson, and Richard Bertram, who explained that with all the liabilities involved, this was a terrible idea. I still think it'd be cool, but my wife thanks them.

I'll always be indebted to the archivists and librarians who fuel nonfiction writing, especially Connie Carter, the patron saint of America's public archives, and Stephanie Marcus at the Library of Congress. Peggy Ann Brown helped me dig, as did Bethany Williams at History Colorado. This work benefited from the work of generations of *Titanic* and wreck scholars. My deepest thanks are to Steve Biel, John Wilson Foster, Steven Spignesi, and the master of the disaster, Walter Lord.

Doug Woolley was one of the richest and kindest subjects I've ever followed. He and his friends—Gary Frank Amphlett Smith, David Simpson, Kamran Ashraf, Rajinder "Tiger" Singh, Adam Magson, and Dolly Brill—helped at every stage of this book. Everyone should want a posse like Doug's. Mine includes Andy Carmona, Liz Flock, Dan Berg, Josh Cohen, Ella Al-Shamahi, and Spencer Millsap. Thanks to Jaclyn Cohen for the "Tea-Tanic" steeper. And Adam Gerber for, among everything else, the title.

I wrote this book while hunkered down during the pandemic thanks to the Georgia hospitality of my world-class in-laws, Julie, Robert, and Winston Ford. My mom, Arlene, is the perennial wind at my back, and my dad, Ron, is my best book marketer. My sister, Karen, is a ray of light bright enough to reach the deepest ocean.

And finally my core: Charlie, the most loyal writing assistant; Alanna, my favorite person to dream with; and Micah, who got here just in time.

# Notes

This is a work of nonfiction, which makes it subject to hard facts and archival sourcing. This is more complicated than it sounds in the world of shipwrecks, a field prone to looping debates, dark mysteries, ceaseless conspiracies, and extreme opinions. What's more, when it comes to older wrecks, deciding whose recollections to believe can be like calling balls and strikes with your eyes closed. Do I trust the account of the person who was *on* the ship, or the scientist who studied the wreck a century later using modern technology? In general, I've included both, but I gave eyewitnesses priority. I also relied heavily on the public record in contemporaneous newspapers and books. Anything within quotes I took from an official source or interview.

## AUTHOR'S NOTE

xii **an incredible three million:** UNESCO, Section of Museums and Cultural Objects, Division of Cultural Objects and Intangible Heritage, *The UNESCO*

*Convention on the Protection of the Underwater Cultural Heritage*, CLT/CIH/MCO/2007/PI/38, UNESCO, 2007.

xii **more than thirty miles per hour:** Walter Lord, *The Night Lives On: New Thoughts, Theories, and Revelations About the Titanic* (London: Penguin Books, 1998).

## PROLOGUE

1 **Cold temperatures produce flakes:** Kenneth G. Libbrecht, "A Guide to Snowflakes," SnowCrystals.com, February 1, 1999, accessed June 28, 2021, http://www.its.caltech.edu/~atomic/snowcrystals/class/class-old.htm.

2 **John Thomas Towson, a scientist:** John Thomas Towson, *Practical Information on the Deviation of the Compass: For the Use of Masters and Mates of Iron Ships* (London: J. D. Potter, 1882).

2 **"iceberg alley":** U.S. Coast Guard, Department of Homeland Security, *International Ice Observation and Ice Patrol in the North Atlantic Ocean* (Washington, DC: Government Printing Office, 2004), 58.

3 **For three years the icy mass:** Alasdair Wilkins, "What Happened to the Iceberg That Sank the *Titanic*?" *Wired*, April 16, 2012, https://www.wired.com/2012/04/titanic-iceberg-history/.

3 **Icebergs had struck ships as long:** Brian T. Hill, "Ship Collision with Iceberg Database," Institute for Ocean Technology, National Research Council Canada (2005), 3.

3 **they grow top-heavy and flip:** Melissa Wiley, "An Iceberg Flipped Over, and Its Underside Is Breathtaking," *Smithsonian Magazine*, January 22, 2015, https://www.smithsonianmag.com/science-nature/photographer-captures-stunning-underside-flipped-iceberg-180953951.

## CHAPTER 1: SHIPFALL

5 **Ocean waves lifted the ship and slammed it on the rocks:** Author unknown, "Account of a Shipwreck, 1693," British Library, https://www.bl.uk/learning/timeline/item104343.html.

5 **"Our ship at last sunk quite down":** "Account of a Shipwreck, 1693."

6 **Chamnan, who for the rest of his life:** Donald Richie, "An Ambassador's Wild Tale of the Wilderness," *Japan Times*, February 22, 2004.

6 **Flooding is the most common reason:** Ajay Menon, "Why Ships Sink—10 Major Reasons," *Marine Insight*, last updated August 25, 2021, https://www.marineinsight.com/naval-architecture/why-ships-sink-10-major-reasons.

6 **"Imagine the headlines if even a single 747":** Susan Casey, *The Wave: In Pursuit of the Rogues, Freaks, and Giants of the Ocean* (New York: Anchor Books, 2011), 11.

7 **Seven Stones Reef:** Richard Larn and Bridget Larn, *Shipwreck Index of the British Isles: East Coast of England: Essex, Suffolk, Norfolk, Lincolnshire,*

*Yorkshire, County Durham, Northumberland* (London: Lloyd's Register of Shipping, 1997).

7 **Kenn Reefs east of Australia:** Jack K. Loney, *Australian Shipwrecks*, vol. 5 (Portarlington, Australia: Marine History Publications, 1986).

7 **rocky straits of Lombok:** "Where Are the Best Places to Go Wreck Diving in Indonesia?," Two Fish Divers, January 31, 2019, https://www.twofishdivers.com/2019/01/wreck-diving-indonesia.

7 **In the span of eight days in August of 2010:** Details of the August 7, 2010, collision of MSC *Chitra* and MV *Khalijia-III* can be found at http://www.fortunes-de-mer.com/old/rubriques/liens%20et%20contacts/detailsactualites/MSC_Chitra_2010.htm.

7 **a delicate name like *Belle Rose*:** Brian Clark Howard, "Photos Reveal Ship Damage to Coral Reef in Shark Sanctuary," *National Geographic*, June 23, 2016, https://www.nationalgeographic.com/science/article/belle-rose-ship-damages-coral-reef-shark-sanctuary-malapascua-philippines.

8 **The ship was so asymmetrically designed:** Greta Franzen, *The Great Ship* Vasa (New York: Hastings House, 1971).

8 **result of liquefaction:** Dave Petley, "Dynamic Liquefaction: A Mass Movement That Can Sink Ships," *The Landslide Blog*, September 3, 2018, accessed June 29, 2021, https://blogs.agu.org/landslideblog/2018/09/03/dynamic-liquefaction-ships.

8 **the *Hui Long*, a midsize cargo vessel:** Michael Munro and Abbas Mohajerani, "Liquefaction Incidents of Mineral Cargoes on Board Bulk Carriers," *Advances in Materials Science and Engineering* (2016).

9 **the SS *City of Glasgow* disappeared:** Colin G. Pooley, *Mobility, Migration and Transport: Historical Perspectives* (Switzerland: Springer International Publishing, 2017).

9 **The SS *Naronic*:** Ronald Rompkey, ed., *Labrador Odyssey: The Journal and Photographs of Eliot Curwen on the Second Voyage of Wilfred Grenfell, 1893* (Montreal: McGill-Queen's University Press, 1999).

9 **Icebergs were such a common scourge:** Brian T. Hill, "Ship Collision with Iceberg Database," Institute for Ocean Technology, National Research Council Canada (2005).

10 **junk ship *Tek Sing*:** Nigel Pickford and Michael Hatcher, *The Legacy of the Tek Sing: China's Titanic—Its Tragedy and Its Treasure* (Cambridge, UK: Granta Editions, 2000).

10 **munitions ship *Mont-Blanc*:** John U. Bacon, *The Great Halifax Explosion: A World War I Story of Treachery, Tragedy, and Extraordinary Heroism* (New York: William Morrow, 2017).

10 ***New York Times* renting out:** Rick Musser, "The 1910s," History of American Journalism, last updated December 31, 2007, accessed June 29, 2021, https://history-journalism.ku.edu/1910/1910.shtml.

11 **Eva Hart:** Eva Hart, as told to Ron Denney, *A Girl Aboard the Titanic* (Gloucestershire, UK: Amberley Publishing, 2014).

11 **"The agony of that night can never be told":** Judith B. Geller, *Titanic: Women and Children First* (London: W. W. Norton, 1998), 120.

12 **voyage data recorders:** Lorenzo Franceschi-Bicchierai, "Pirate Hackers Can Easily Spy on Ships Through Insecure 'Black Boxes,'" *Vice*, December 9, 2015, https://www.vice.com/en/article/53dzn5/pirate-hackers-can-easily-spy-on-ships-through-insecure-black-boxes.

13 **nautical forensics by *National Geographic*:** TITANIC EXPLORA, "How Sank the TITANIC? New Sinking Theory 2012 by James Cameron," December 5, 2012, YouTube video, 2:57, https://www.youtube.com/watch?v=YyYMY7JsNNY.

14 **length of rope a mile deep:** Peter J. Herring and Malcolm R. Clarke, eds., *Deep Oceans* (London: Praeger, 1971).

14 **an implosion so forceful:** Walter Lord, *A Night to Remember: The Classic Account of the Final Hours of the Titanic* (1955; New York: Henry Holt and Company, 2005).

14 **Gabr scuba dove:** Elizabeth Palermo, "Man Sets World Record for Deepest Underwater Dive," *Scientific American*, September 26, 2014.

15 **water below six hundred feet turns black:** "How Far Does Light Travel in the Ocean?," National Ocean Service, last updated November 5, 2021, https://oceanservice.noaa.gov/facts/light_travel.html.

15 **twilight zone:** "Exploring Our Fluid Earth," University of Hawai'i, accessed February 16, 2020, https://manoa.hawaii.edu/ExploringOurFluidEarth.

15 **fish, mollusks, crustaceans:** "The Open Ocean," MarineBio Conservation Society, published June 17, 2018, https://www.marinebio.org/oceans/open-ocean.

15 **chemical bioluminescence:** "Bioluminescence," Smithsonian Ocean—Find Your Blue, December 18, 2018, https://ocean.si.edu/ocean-life/fish/bioluminescence.

15 **tubeshoulders:** Joseph S. Nelson, Terry C. Grande, and Mark V. H. Wilson, *Fishes of the World*, 5th ed. (New York: Wiley, 2016).

16 **excavate the ship's famous telegraph:** Vanessa Romo, "Expedition to Salvage *Titanic*'s Wireless Telegraph Gets the Go-Ahead," NPR, May 20, 2020, https://www.npr.org/2020/05/20/859960432/expedition-to-salvage-titanics-wireless-telegraph-gets-the-go-ahead.

17 **the ship's front hatch cover:** Maryanne Culpepper, "*Titanic*: The Final Word with James Cameron," National Geographic TV, Vimeo clip, 1:38:52, https://vimeo.com/87420251.

18 **women and children go last:** Mikael Elinder and Oscar Erixson, "Gender, Social Norms, and Survival in Maritime Disasters," *Proceedings of the*

*National Academy of Sciences of the United States of America* 109, no. 33 (2012): 13220–3224, doi:10.1073/pnas.1207156109.

18 **argument against women's suffrage:** Alison George, "Sinking the Titanic 'Women and Children First' Myth," *New Scientist*, July 30, 2012, https://www.newscientist.com/article/dn22119-sinking-the-titanic-women-and-children-first-myth.

19 **an investigation by the British Board of Trade:** Brian J. Ticehurst, "British Wreck Commissioner's Inquiry," Titanic Inquiry Project, accessed June 12, 2020, https://www.titanicinquiry.org/BOTInq/BOT01.php.

19 **a fashion designer named Lucy Duff-Gordon:** Lucy Duff-Gordon, *Discretions and Indiscretions* (New York: Frederick A. Stokes Company, 1932).

19 **ships in distress would use SOS:** *Radio News* (New York: Experimenter Publishing Company, 1923), 435.

20 **The technology had its limits:** Sungook Hong, "Marconi's Error: The First Transatlantic Wireless Telegraphy in 1901," *Social Research* 72, no. 1 (2005): 107–24.

20 **communicate was by colored flags:** *The 1931 International Code of Signals* (Washington, DC: Government Printing Office, 1933).

20 **passengers were permitted:** Erin Blakemore, "Why *Titanic*'s First Call for Help Wasn't an SOS Signal," *National Geographic*, May 28, 2020, https://www.nationalgeographic.com/history/article/why-titanic-first-call-help-not-sos-signal.

21 **city editor named Charles Crane:** "Four-Word Bulletin Touched Off Frenzy," *The Columbus Dispatch*, April 14, 2012.

## CHAPTER 2: THE DEATH AND BIRTH OF GREAT SHIPS

25 **One summer when he was a boy:** Doug Woolley, in-person interview by author, London, March 6, 2020.

25 **"At this rate, you'll never":** Woolley, in-person interview by author.

26 **"The characters I created":** Doug Woolley, in-person interview by author, London, March 7, 2020.

27 **Davy Jones's locker:** Tobias George Smollett, *The Adventures of Peregrine Pickle* (London: R. Baldwin and J. Richardson; and D. Wilson and T. Durham, 1758).

27 **British scientist named William Keatinge:** W. R. Keatinge, "Death After Shipwreck," *The British Medical Journal* 2, no. 5477 (1965): 1537–41.

28 **Hypothermia happens twenty-five times faster:** "Cold Stress—Cold Water Immersion," Centers for Disease Control and Prevention, last updated June 6, 2018, https://www.cdc.gov/niosh/topics/coldstress/coldwaterimmersion.html.

30 **World War II wrecks:** "A Regional Strategy to Address Marine Pollution from World War II Wrecks," *Secretariat of the Pacific Regional Environment*

*Programme*, July 2002, https://www.sprep.org/attachments/Legal/Marinewrecks.pdf.

30 **desert of Namibia:** "Three Shipwrecks on Namibia's Skeleton Coast," Namibia: Endless Horizons, accessed March 22, 2020, https://namibiatourism.com.na/blog/Three-Shipwrecks-on-Namibia-s-Skeleton-Coast.

30 **cornfields in Kansas:** Luke Spencer, "How a Champagne-Laden Steamship Ended Up in a Kansas Cornfield," *Atlas Obscura*, July 20, 2016, https://www.atlasobscura.com/articles/how-a-champagneladen-steamship-ended-up-in-a-kansas-cornfield.

30 **After the Twin Towers fell:** Kate Langin, "Wooden Ship Unearthed at World Trade Center Site from Revolutionary-Era Philadelphia," *National Geographic*, July 31, 2014, https://www.nationalgeographic.com/history/article/140731-world-trade-center-ship-tree-rings-science-archaeology.

30 **$60 billion:** Rob Goodier, "What's the Total Value of the World's Sunken Treasure?," *Popular Mechanics*, February 22, 2012, https://www.popularmechanics.com/technology/infrastructure/a7425/whats-the-total-value-of-the-worlds-sunken-treasure.

31 **"the most valuable treasure":** "'The Most Valuable Treasure in History': Colombian President Declares Shipwreck Has Lots of Loot but Location Is Being Kept Top Secret," Associated Press, December 6, 2015.

31 **In 2016, divers working for the Dutch government:** Kathryn Miles, "The Thieves Who Steal Sunken Warships, Right Down to the Bolts," *Outside*, May 2, 2017.

33 **the USS *Cole*:** *Lessons Learned from the Attack on U.S.S.* Cole, *on the Report of the Crouch-Gehman Commission, and on the Navy's Judge Advocate General Manual Investigation into the Attack, Including a Review of Appropriate Standards of Accountability for U.S. Military Services*, 107th Cong. (2001).

34 **Henry Ford to the White House:** *Dictionary of American Naval Fighting Ships: Historical Sketches: Letters R through S. Appendices: Submarine Chasers, Eagle-Class Patrol Craft* (Washington, DC: Government Printing Office, 1959), 744.

34 **construction began on the *Titanic*:** William MacQuitty, *Titanic Memories: The Making of "A Night to Remember"* (London: National Maritime Museum, 2000).

35 **No. 3 bar iron instead of the top-of-the-line No. 4 bar:** William Broad, "In Weak Rivets, a Possible Key to *Titanic*'s Doom," *The New York Times*, April 15, 2008.

35 **metallurgical analysis:** Tim Foecke, "Metallurgy of the RMS *Titanic*," Gaithersburg, MD: U.S. Department of Commerce, Technology Administration, National Institute of Standards and Technology, Materials Science and Engineering Laboratory, 1998.

36 **"Being in a ship":** *Pacific Marine Review* 44 (1947): 37.

36 **In 1797, the *Parr*:** Brendan Wolfe, "Slave Ships and the Middle Passage," *Encyclopedia Virginia*, February 1, 2021.

36 **"spoonways":** *Liverpool and Slavery: An Historical Account of the Liverpool-African Slave Trade* (Liverpool, UK: A. Bowker & Son, 1884).

37 **An anonymously published account:** *Liverpool and Slavery.*

37 **As many as one in six:** Elizabeth Mancke and Carole Shammas, eds., *The Creation of the British Atlantic World* (Baltimore: Johns Hopkins University Press, 2015).

38 **Revenue Cutter Service:** William H. Thiesen, "The Long Blue Line: Merry Christmas Coast Guard—180 Years of Search and Rescue!," Coast Guard Compass, December 21, 2017, https://coastguardblogcom.wpcomstaging.com/2017/12/21/tlbl-180-years-search-and-rescue.

38 **In January 1837, a three-masted sailing barque:** Thiesen, "The Long Blue Line."

38 **the *Savannah*, in 1819:** Bradley Sheard, *Lost Voyages: Two Centuries of Shipwrecks in the Approaches to New York* (New York: Aqua Quest Publications, 1998).

39 **paddle steamer *Great Western*:** Francis Bradley, International Marine Engineering (New York: Simmons-Boardman Publishing Company, 1910), 417.

39 **Nearly all advances in shipping:** Paul Heyer, *Titanic Legacy: Disaster as Media Event and Myth* (New York: Praeger, 1995).

39 **American-based Collins line:** "The Steamships of the Collins Line," *Scientific American*, April 24, 1858.

39 ***Great Eastern*, a nineteen-thousand-ton leviathan:** G. Hutchinson, *History of the "Great Eastern" Steamship* (Liverpool, UK: Cassell, 1887).

41 **hobbled the White Star Line:** Wilton J. Oldham, *The Ismay Line: The Titanic, the White Star Line and the Ismay Family* (Luton, UK: Andrews UK Limited, 2013).

41 **survivors who had been picked up:** Charles R. Pellegrino, *Ghosts of the Titanic* (New York: HarperCollins, 2000).

42 **about 60 decibels:** "What Noises Cause Hearing Loss?," Centers for Disease Control and Prevention, October 7, 2019, https://www.cdc.gov/nceh/hearing_loss/what_noises_cause_hearing_loss.html.

42 **Blue whale calls:** Ella Davies, "The World's Loudest Animal Might Surprise You," *Asia One*, April 8, 2016, https://www.asiaone.com/worlds-loudest-animal-might-surprise-you.

42 **In the 1940s, marine scientists:** National Research Council of the National Academies, *Ocean Noise and Marine Mammals* (Washington, DC: National Academies Press, 2003).

43 **geophysicist named Maurice Ewing:** Paul Théberge, Kyle Devine, and Tom Everrett, eds., *Living Stereo: Histories and Cultures of Multichannel Sound* (New York: Bloomsbury Academic, 2015).

43 **sound fixing and ranging (or SOFAR) channel:** "What Is SOFAR?," National Ocean Service, last updated February 26, 2021, https://oceanservice.noaa.gov/facts/sofar.html.

## CHAPTER 3: THE MOVEMENT FROM ORDER TO CHAOS

45 **Late Heavy Bombardment:** Adam Mann, "Bashing Holes in the Tale of Earth's Troubled Youth," *Nature* 553, no. 7689 (2018): 393–95, doi:10.1038/d41586-018-01074-6.

46 **The second theory:** Linda T. Elkins-Tanton, "Formation of Early Water Oceans on Rocky Planets," *Astrophysics and Space Science* 332, no. 2 (2010): 359–64, doi:10.1007/s10509-010-0535-3.

46 **"I *love* talking about this":** Elkins-Tanton, "Formation of Early Water Oceans on Rocky Planets"; Linda T. Elkins-Tanton, telephone interview by author, May 6, 2020.

46 **Inside every rock:** Brandon Schmandt et al., "Dehydration Melting at the Top of the Lower Mantle," *Science* 344, no. 6189 (2014): 1265–68, doi:10.1126/science.1253358.

47 **all of the water on earth:** "Where Is Earth's Water?," U.S. Geological Survey, October 25, 2019, https://www.usgs.gov/media/images/distribution-water-and-above-earth.

49 **require Congress to do something:** Tom Kuntz, *The Titanic Disaster Hearings: The Official Transcripts of the 1912 Senate Investigation* (New York: Pocket Books, 1998).

49 **Senator Smith was less interested:** Daniel Allen Butler, *The Other Side of the Night: The Carpathia, the Californian, and the Night the Titanic Was Lost* (Philadelphia: Casemate Publishers, 2009), 139.

50 **the Waldorf-Astoria:** Butler, *The Other Side of the Night*, 147.

52 **"What were the circumstances":** Kuntz, *The Titanic Disaster Hearings*, 14.

53 **The British Board of Trade conducted:** John Charles Bigham Mersey and Hon. Arthur Gough-Calthorpe, *Loss of the Steamship "Titanic": Report of a Formal Investigation into the Circumstances Attending the Foundering on April 15, 1912, of the British Steamship "Titanic," of Liverpool* (Washington, DC: Government Printing Office, 1912).

53 **proposed regulations:** John Lang, *Titanic: A Fresh Look at the Evidence by a Former Chief Inspector of Marine Accidents* (Lanham, MD: Rowman & Littlefield Publishers, 2012).

54 **"This should be the occasion":** "United States Senate Inquiry," Titanic Inquiry Project, accessed May 2020, https://www.titanicinquiry.org/USInq/USReport/AmInqRepSmith02.php.

54 **"it marks the movement":** Josiah Blackmore, *Manifest Perdition: Shipwreck Narrative and the Disruption of Empire* (Minneapolis: University of Minnesota Press, 2002), 52.

55 **SS *Atlantus*, an old concrete ship:** "The Concrete Ship SS *Atlantus*," *Cape May County Herald*, August 18, 2017, https://www.capemaycountyherald.com/community/article_0fd91156-844c-11e7-9a9d-0b206a7982c0.html.

55 **"I've been trying to get":** Parks Stephenson, telephone interview by author, May 21, 2020.

56 **sitting awkwardly upside down:** Sam Roberts, "99 Years Later, Navy Probing Warship Disaster Off Long Island," *The New York Times*, September 17, 2017.

56 **high density of wrecks:** Fiona Young-Brown, "The Place Where 1,000 Ships Were Sunk," *BBC Earth*, November 21, 2016.

57 **Kruszelnicki published a study:** Adam Lusher, "Scientist 'Solves' Mystery of the Bermuda Triangle—by Claiming There Was No Mystery in the First Place," *The Independent*, July 27, 2017.

57 **"To say quite a few ships and airplanes":** Brian Clark Howard, "Gas Craters Off Norway Linked to Fringe Bermuda Triangle Theory," *National Geographic*, March 15, 2016, https://www.nationalgeographic.com/culture/article/160315-norway-craters-methane-hydrates-bermuda-triangle-science.

57 **USS *Cyclops* and USS *Proteus*:** Kathryn Walker, *Mysteries of the Bermuda Triangle* (London: Crabtree Publishing Company, 2008), 12.

57 **gas bubbles from deep-sea methane vents:** "Do Giant Gas Bubbles Explain the Mystery of the Bermuda Triangle?," *The Guardian*, March 14, 2016.

58 **German naval commanders decided to scuttle:** "The Scuttling of the German High Seas Fleet at Scapa Flow," Naval History and Heritage Command, September 10, 2019, https://www.history.navy.mil/browse-by-topic/wars-conflicts-and-operations/world-war-i/history/scapa-flow.html.

58 **cleaned out the *Ford*:** "LCS Fires Ship-Killing Missile During Ex-Frigate Ford's SINKEX," *The Navy Times*, October 2, 2019.

59 **Operation CHASE:** William R. Brankowitz, "Chemical Weapons Movement History Compilation," Department of Defense, Aberdeen Proving Ground, MD, Office of the Program Manager for Chemical Munitions (Demilitarization and Binary) (Provisional), 1987.

60 **concentrated mustard gas:** Russell Fendick et al., "Notes from the Field: Exposures to Discarded Sulfur Mustard Munitions—Mid-Atlantic and New England States 2004–2012," *Morbidity and Mortality Weekly Report*, April 26, 2013.

60 **In 1987, hundreds of dolphins:** John R. Bull, "The Deadliness Below," *Daily Press* (Newport News, VA), October 30, 2005.

61 **the scuttling operation is believed to have included:** Paul E. Rosenfeld and Lydia G. H. Feng, *Risks of Hazardous Wastes* (Amsterdam: Elsevier Science, 2011), 51.

61 **"It's impossible to know":** Craig Williams, telephone interview by author, March 16, 2021.

62 **"On account of the weight":** "How the *Titanic* Looks Now," *The Medford Patriot* (Oklahoma), May 16, 1912.

62 **Maury was one of the first:** Matthew Fontaine Maury, *The Physical Geography of the Sea, and Its Meteorology* (New York: Harper & Brothers, 1875).

64 **"It sometimes happens that":** "The *Titanic*'s Grave on the Ocean Floor," *The Kansas City Star*, April 21, 1912.

64 **deep-sea creatures evolved:** David J. Randall and Anthony P. Farrell, eds., *Deep-Sea Fishes* (San Diego, CA: Academic Press, 1997).

64 **Trinity Bay in 1877:** "Look Out for the Squid," *The St. Albans Daily Messenger* (Vermont), July 6, 1897.

64-65 **scientists filmed a giant squid:** Brigit Katz, "Watch First Footage of Giant Squid Filmed in American Waters," *Smithsonian Magazine*, June 24, 2019, https://www.smithsonianmag.com/smart-news/first-time-giant-squid-was-filmed-american-waters-180972479.

65 **But every culture:** Nathaniel Altman, *The Deva Handbook: How to Work with Nature's Subtle Energies* (Rochester, VT: Destiny Books, 1995), 23.

66 **U.S. Navy spends $3 billion:** Kyle Mizokami, "Rust: The U.S. Navy's $3 Billion-a-Year Oxidation Problem," *Popular Mechanics*, January 15, 2020, https://www.popularmechanics.com/military/navy-ships/a30522792/navy-fighting-rust.

66 **"Most calls I get":** Alana Fossa, telephone interview by author, May 12, 2020.

68 **The *L.R. Doty*, a Gilded Age cargo steamer:** Brendon Baillod, "Steamer *L.R. Doty* Found in Lake Michigan Off Milwaukee," *Shipwreck World*, June 23, 2010, https://www.shipwreckworld.com/articles/steamer-doty-shipwreck.

68 **age bottles underwater:** Christopher Osburn, "Why One Napa Winery Is Experimenting with Aquaoir, Wine Aged Under Water," *Eater*, August 21, 2015, https://www.eater.com/drinks/2015/8/21/9184763/aquaoir-wine-aging.

68 **the Italian troopship *Umbria*:** "Dive One of the World's Best Wrecks, the *Umbria* in Sudan," *AquaViews Online Scuba Magazine*, June 25, 2017, https://www.leisurepro.com/blog/scuba-dive-destinations/dive-worlds-wrecks-umbria-sudan.

68 **found the oldest shipwreck:** Nick Romeo, "Centuries of Preserved Shipwrecks Found in the Black Sea," *National Geographic*, October 26, 2016, https://www.nationalgeographic.com/history/article/black-sea-shipwreck-discovery.

69 **abandoned wood hulls for steel ones:** Erich W. Zimmermann, *Modern Business: Foreign Trade and Shipping* (New York: Alexander Hamilton Institute, 1919), 223.

## CHAPTER 4: MERELY A MATTER OF MAGNETS

71 **The earliest sailors:** Brian Lavery, *A Short History of Seafaring* (London: Dorling Kindersley Limited, 2013).

72 **embarrassing wreck of the HMS *Royal George*:** Hilary L. Rubinstein, *Catastrophe at Spithead: The Sinking of the Royal George* (Barnsley, UK: Pen & Sword Books, 2020).

73 **whaling ship *Essex*:** Nathaniel Philbrick, *In the Heart of the Sea: The Tragedy of the Whaleship* Essex (New York: Penguin Books, 2015).

73 **Custom of the Sea:** A. W. Brian Simpson, *Cannibalism and the Common Law* (Chicago: University of Chicago Press, 1984).

73 **"the most stupendous vessel":** "City Intelligence," *The Evening Post* (New York), January 28, 1850.

74 **Smith had been born in 1861:** Wilbur Fiske Stone, *History of Colorado*, vol. 2 (Chicago: S. G. Clarke Pub., 1918).

75 **"My object, first of all":** "Giant Magnets to Raise the *Titanic*," *The Pittsburgh Press*, March 8, 1914.

76 **Smith's view of finding the ship:** "Giant Magnets to Raise the Sunken *Titanic*," *Daily Arkansas Gazette* (Little Rock), March 8, 1914.

79 **"In three months":** "Giant Magnets to Raise the Sunken *Titanic*."

79 **"It is merely":** "Giant Magnets to Raise the Sunken *Titanic*."

80 **Benjamin Franklin studied the Gulf Stream:** Deborah Heiligman, *The Mysterious Ocean Highway: Benjamin Franklin and the Gulf Stream* (Chicago: Heinemann Raintree, 1999).

80 **Riker released:** "To Move the Earth and Melt the Pole," *The New York Times*, September 29, 1912.

80 **He proposed a two-hundred-mile-long rock jetty:** Uwe Kitzinger and Ernst G. Frankel, *Macro-Engineering and the Earth: World Projects for Year 2000 and Beyond* (West Sussex, UK: Horwood Publishing, 1998).

81 **"I hear the exclamation 'visionary'":** "To Move the Earth and Melt the Pole," *The Houston Post*, October 6, 1912.

81 **"Man Can Control All":** Kitzinger and Frankel, *Macro-Engineering and the Earth.*

81 **sound navigation ranging, also known as sonar:** Author unknown, *Think* (Endicott, NY: International Business Machines Corporation, 1958), 36.

82 **Pino boarded his craft:** "A Rival to Marconi," *The Indianapolis Journal*, February 8, 1903.

82 ***Madagascar*, a British frigate:** Lurline Stuart, "The Lost Gold Ship," *The LaTrobe Journal* 67 (2001), http://latrobejournal.slv.vic.gov.au/latrobejournal/issue/latrobe-67/t1-g-t2.html.

82 **even the *Mary Celeste*:** Valerie Martin, *The Ghost of the* Mary Celeste (London: Knopf Doubleday Publishing Group, 2015).

83 **a basic napkin sketch:** "Hopes to Raise the *Titanic,*" *The Ordway New Era* (Colorado), April 17, 1914.

84 **salvage underwater vessels:** Charles A. Bartholomew and William I. Milwee, *Mud, Muscle, and Miracles: Marine Salvage in the United States Navy* (Washington, DC: U.S. Department of the Navy, 1990).

84 **marine insurance companies:** National Ocean Industries Association and NOAA, *Where Land and Water Meet* (Washington, DC: U.S. Department of Commerce, 1979), 45.

85 **the *Isla de Luzón*, the *Isla de Cuba*, and the *Don Juan D'Austria*:** Bartholomew and Milwee, *Mud, Muscle, and Miracles.*

85 **the submarine *F-4*:** *The Submarine in the United States Navy* (Washington, DC: Government Printing Office, 1963).

87 **combined forces in 1910 to refloat the *Maine*:** *Raising the Wreck of the United States Battleship Maine: Havana Harbor* (Buffalo, NY: Lackawanna Steel Company, 1912).

## CHAPTER 5: LUNGS THE SIZE OF ACORNS

92 **Smith needed, he raised:** Charles Smith, "A Plan to Lift the *Titanic,*" unpublished typescript, University of Colorado, 1914.

92 **the average American salary:** Witt Bowden, "War and Postwar Wages, Prices, and Hours, 1914–23 and 1939–44," bulletin no. 852 (1945).

93 **"ridiculous proposal":** "Engineering: A Ridiculous Proposal," *Scientific American*, February 14, 1914.

94 **"the Gulf Stream and the Labrador current":** "Engineer Hopes to Raise *Titanic* with Electric Magnets, Submarines and Floating Camels," *Evening Star* (Washington, DC), March 8, 1914.

94 **in Long Island Sound:** James L. Mooney, *Dictionary of American Naval Fighting Ships,* vol. 3 (Washington, DC: Government Printing Office, 1968).

94 **Darwin seemed to believe:** Charles Darwin, *The Origin of Species* (New York: P. F. Collier, 1909).

95 **Lake taunted Smith:** "A Denver Engineer Hopes to Raise the *Titanic,*" *Buffalo Sunday Morning News*, March 15, 1914.

95 **"If you are at all familiar":** "Hopes to Raise the *Titanic,*" *Nashville Journal*, April 16, 1914.

97 **Armstrong got suddenly hot:** Kim Bowden, "Armstrong a True Pioneer of Aviation Medicine and Fitting USAFSAM Exemplar," Wright-Patterson Air Force Base, March 26, 2018, https://www.wpafb.af.mil/News/Article-Display/Article/1472150/armstrong-a-true-pioneer-of-aviation-medicine-and-fitting-usafsam-exemplar.

98 **how to make artificial pressures:** Kenneth Chang, "The Big Squeeze," *The New York Times*, December 16, 2013, https://www.nytimes.com/2013/12/17/science/the-big-squeeze.html.

98 **peanut butter turns to diamonds:** Robert M. Hazen, *The Diamond Makers* (Cambridge, UK: Cambridge University Press, 2000).

98 **"The wife of a diver, poor woman!":** Cleveland Moffett, *St. Nicholas: An Illustrated Magazine for Young Folks* (New York: The Century Co., 1901), 207.

99 **"I got into the suit":** Moffett, *St. Nicholas: An Illustrated Magazine for Young Folks*, 208.

100 **"A western genius":** "A Western Genius Declares," *Escanaba Morning Press* (Michigan), February 6, 1914.

100 **"Personally we have no inclination":** A. V. Napler, "Reflections and Remarks," *Iola Register* (Kansas), February 16, 1914.

100 **the USS *Saint Paul*:** S. Swiggum and M. Kohli, "Ship Descriptions," *The Ship List*, last updated October 3, 2016, http://www.theshipslist.com/ships/descriptions/ShipsS.shtml.

100 ***Seawise Giant* (and later renamed *Knock Nevis*):** Ethan Trex, "*Seawise Giant*: You Can't Keep a Good Ship Down," *Mental Floss*, June 1, 2011, https://www.mentalfloss.com/article/27877/seawise-giant-you-cant-keep-good-ship-down.

101 ***Per Brahe* grew into a *Titanic*-like fable:** August Håkansson, "Shipwreck Skeleton May Be Swedish Captain," *The Local*, October 26, 2015, https://www.thelocal.se/20151026/shipwreck-skeleton-may-be-swedish-captain.

101 ***Per Brahe* was raised in 1922:** *Perilous Passage*, directed by Thom Britten-Austin and Brent Johansson (Sweden, 2020), https://en.filmenomperbrahe.com/.

102 **Smith sought amounted to:** Christopher Chantrill, "Federal 1914 Government Spending," U.S. Government Spending, accessed June 11, 2020, https://www.usgovernmentspending.com/fed_spending_1914USrn.

103 **the *Carib* and the *Evelyn*:** "No Protest Will Be Made for Sinking *Evelyn*," *Riverside Daily Press*, February 22, 1915.

104 ***Flaba* was sunk by a U-boat:** Frank Blazich Jr., "United States Navy and World War I: 1914–1922," Naval History and Heritage Command, June 10, 2020, https://www.history.navy.mil/research/library/online-reading-room/title-list-alphabetically/u/us-navy-world-war-i-redirect.html.

104 **British foreign secretary admitted:** Alan Travis, "*Lusitania* Divers Warned of Danger from War Munitions in 1982, Papers Reveal," *The Guardian*, April 30, 2014.

105 **The price of zinc:** G. F. Loughlin, *The Oxidized Zinc Ores of Leadville Colorado* (Washington, DC: Government Printing Office, 1918).

105-106 **Interocean Submarine Engineering Company:** "Hand of Wall Street Reaches Out After Sunken Treasures of the Deep," *The Philadelphia Inquirer*, June 4, 1916.

106 **Italian troopship, the *Principe Umberto*:** "Italian Transport Sunk by Torpedo," *The New York Times*, June 11, 1916.

## CHAPTER 6: I REGARD THE *TITANIC* AS MINE

107 **Beesley was caught up in the excitement:** William MacQuitty, *Titanic Memories: The Making of "A Night to Remember"* (London: National Maritime Museum, 2000).

107 **Beesley, from Derbyshire in the East Midlands:** Lawrence Beesley, *The Loss of the S.S. Titanic: Its Story and Its Lessons* (Boston and New York: Houghton Mifflin, 1912).

109 **"Never have I experienced":** MacQuitty, *Titanic Memories*, 14.

110 **"needed no added drama":** MacQuitty, *Titanic Memories*, 23.

110 **Doug Woolley saw:** Doug Woolley, in-person interview by author, London, March 7, 2020.

111 **the *Empress* sank:** Anne Renaud, *Into the Mist: The Story of the* Empress of Ireland (Toronto: Dundurn Press, 2010).

111 ***Eastland* sank:** Susan Q. Stranahan, "The *Eastland* Disaster Killed More Passengers Than the *Titanic* and the *Lusitania*. Why Has It Been Forgotten?," *Smithsonian Magazine*, October 27, 2014.

111 **Eight hundred forty-five:** Jay R. Bonansinga, *The Sinking of the* Eastland*: America's Forgotten Tragedy* (New York: Citadel Press/Kensington Publishing, 2004).

112 **books, newspapers, and at least one early film:** Stephen Bottomore, *The Titanic and Silent Cinema* (East Sussex, UK: Projection Box, 2000).

112 **Each April, sailors and seamen:** Steven Biel, *Down with the Old Canoe: A Cultural History of the Titanic Disaster* (London: W. W. Norton & Company, 1997).

113 **the Shine story:** Paul Heyer, *Titanic Century: Media, Myth, and the Making of a Cultural Icon* (Santa Barbara, CA: Praeger, 2012).

114 **"Had the *Titanic* been a mudscow":** "The *Titanic* Tragedy," *Appeal to Reason*, May 4, 1912.

114 **half a million people died each year:** Steven Biel, ed., *American Disasters* (London: New York University Press, 2001), 322.

115 **Doug's eccentric personality:** Clive Amphlett and Douglas John Faulkner-Woolley, *Titanic: One Man's Dream: Douglas John Faulkner-Woolley: His Claims on Britain's Two Most Famous Liners (QE1 and Titanic): A Biography* (Ilford, UK: Seawise Publication, 1998).

116 **But it also lit a fire in him:** Doug Woolley, in-person interview by author, London, March 7, 2020.

116 **Titanic Enthusiasts of America:** "The *Titanic* Commutator—Articles About the *Titanic*," Titanic Historical Society Inc., accessed August 18, 2020, https://titanichistoricalsociety.org/titanic-commutator.

117 **Lloyd's of London underwriters:** Stephen W. Hines, *Titanic: One Newspaper, Seven Days, and the Truth That Shocked the World* (Naperville, IL: Sourcebooks, 2011).

117 **Lloyd's were eager to disassociate:** "Lloyd's and the *Titanic*," Lloyd's of London, accessed June 11, 2020, https://www.lloyds.com/about-lloyds/history/catastophes-and-claims/titanic.

118 **"Ninety-nine percent of the time":** Nick Gaskell, telephone interview by author, June 17, 2020.

119 **the SS *Central America*:** Gary Kinder, *Ship of Gold in the Deep Blue Sea: The History and Discovery of the World's Richest Shipwreck* (New York: Grove/Atlantic, 2009).

119 **"It saddens me":** "Insurers to Share Shipwreck's Gold," *The New York Times*, August 28, 1992.

122 **"I regard the *Titanic*":** Arthur Veysey, "Report from Europe: British Treasure Seekers Plan to Raise 2 Liners," *Chicago Tribune*, November 2, 1963.

122 **"I'm the only person to have filed":** "Salvage Men Try for the Miracle of the Century," *Aberdeen Press and Journal*, March 20, 1972.

123 **the device, called the bathysphere:** William Beebe, *Half Mile Down* (New York: Harcourt Brace and Company, 1934).

123 **perfected an underwater breathing device:** Jennifer Berne, *Manfish: A Story of Jacques Cousteau* (San Francisco, CA: Chronicle Books, 2012).

124 **known as the Challenger Deep:** Ben Taub, "Thirty-Six Thousand Feet Under the Sea," *The New Yorker*, May 18, 2020.

124 **"It was like looking into a bowl of milk":** Taub, "Thirty-Six Thousand Feet Under the Sea."

124 **Tharp studied sonar profiles:** Hali Felt, *Soundings: The Story of the Remarkable Woman Who Mapped the Ocean Floor* (New York: Henry Holt, 2013).

125 **Beazley started demolishing ships:** Roy Martin and Lyle Craigie-Halkett, *Risdon Beazley: Marine Salvor* (Southampton, UK: Brook House, 2007).

126 **Beazley told no one:** John P. Eaton and Charles A. Haas, *Titanic: Triumph and Tragedy* (London: W. W. Norton, 1995), 301.

126 **a stretch of shore known as the Treasure Coast:** Stephen M. Voynick, *The Mid-Atlantic Treasure Coast: Coin Beaches & Treasure Shipwrecks from Long Island to the Eastern Shore* (Wilmington, DE: Middle Atlantic Press, 1984).

127 ***Today's the day*:** Eric Pace, "Mel Fisher, 76, a Treasure Hunter Who Got Rich Undersea," *The New York Times*, December 21, 1998.

## CHAPTER 7: BATHTUB EXPERIMENTS

130 **Woolley's original plan:** Clive Amphlett and Douglas John Faulkner-Woolley, *Titanic: One Man's Dream: Douglas John Faulkner-Woolley: His Claims on Britain's Two Most Famous Liners (QE1 and Titanic): A Biography* (Ilford, UK: Seawise Publication, 1998).

130 **Woolley revealed this plan in a press conference:** Doug Woolley, telephone interview by author, June 21, 2020.

132 **"gone international":** Doug Woolley, telephone interview by author, April 17, 2020.

132 **New York con artist George Parker:** Craig McGuire, *Brooklyn's Most Wanted: The Top 100 Criminals, Crooks and Creeps from the County of the Kings* (Denver, CO: WildBlue Press, 2017).

132 **Every deck on a ship has a name:** U.S. Navy, *Ship Shapes: Anatomy and Types of Naval Vessels* (Washington, DC: Government Printing Office, 1942).

133 **The *Titanic* was built with ten decks:** Philip Wilkinson, *Titanic: Disaster at Sea* (Mankato, MN: Capstone Press, 2012).

134 **the SS *President Coolidge*, an art deco passenger liner:** Robby Myers, "History of Vanuatu's *President Coolidge* Shipwreck," *Scuba Diving*, November 20, 2017, https://www.scubadiving.com/history-vanuatu-president-coolidge-shipwreck#page-8.

135 **"One way to think":** Ian MacLeod, telephone interview by author, April 30, 2020.

135 **the SS *Great Britain*:** Helen Doe, *SS Great Britain: Brunel's Ship, Her Voyages, Passengers and Crew* (Gloucestershire, UK: Amberley Publishing, 2019).

136 **It was showered with rose petals:** Andrew Bomford, "SS *Great Britain*: From Seabed to National Treasure," *BBC News*, July 4, 2010, https://www.bbc.com/news/10490928.

136 **engineers constructed a giant dehumidification chamber:** Maria Burke, "Saving a Steam Ship," *Chemistry World*, July 1, 2005, https://www.chemistryworld.com/features/saving-a-steam-ship/3004772.article.

137 **the conservation took three years:** "Brunel's Pride Restored," *The Telegraph*, January 6, 2005.

138 **"Titanic Task to Stop Clock":** "Titanic Task to Stop Clock," *The Ottawa Citizen*, November 18, 1968.

139 **"The lower depth the less oxygen . . .":** "Raise the *Titanic*?," *The Province* (Vancouver, BC), August 13, 1968.

139 **Oxygen is at a maximum at the surface:** A. Paulmier and D. Ruiz-Pino, "Oxygen Minimum Zones (OMZs) in the Modern Ocean," *Progress in Oceanography* 80, nos. 3–4 (2008): 113–28, doi:10.1016/j.pocean.2008.08.001.

139 **Woolley worked on this experiment in his bathtub:** Doug Woolley, telephone interview by author, August 16, 2020.

140 **"Of course I know it's an odd obsession":** Colin Smith, "Three-Nation Team Will Try to Raise *Titanic*," *The Sacramento Bee*, October 17, 1969.

140 **Woolley never met anyone on his team:** "11-Man Team Plans *Titanic* Raising," *The Tampa Times*, October 17, 1969.

140 **the £4.8 million someone had suggested:** "Question: Who Owns Lost Ship?," *Des Moines Tribune*, October 20, 1969.

140 **every member of the team:** Doug Woolley, telephone interview by author, August 17, 2020.

142 **Project Azorian, as the top secret maneuver:** Norman C. Polmar and Michael White, *Project Azorian: The CIA and the Raising of the* K-129 (Annapolis, MD: Naval Institute Press, 2010).

142 **The *Glomar* included:** "Project AZORIAN," Central Intelligence Agency, accessed September 3, 2020, https://www.cia.gov/legacy/museum/exhibit/project-azorian.

143 **"What are you doing here?":** Polmar and White, *Project Azorian*, 111.

144 **Jack Anderson wasn't convinced:** Josh Dean, *The Taking of K-129: How the CIA Used Howard Hughes to Steal a Russian Sub in the Most Daring Covert Operation in History* (New York: Dutton Caliber, 2017).

145 **a four-hundred-twenty-square-mile area:** Charles Fishman, "The Epic Search for the *Challenger*," *The Washington Post*, May 28, 1986, https://www.washingtonpost.com/archive/lifestyle/1986/05/28/the-epic-search-for-the-challenger/2c445a93-f39c-448b-b4c0-ca79efe3ad48.

145 **The navy provided an NR-1:** "Report on the Salvage of the Space Shuttle *Challenger* Wreckage," U.S. Navy Naval Seal Systems Command, 1988.

146 **an unmanned vehicle discovered:** Penny Pagano, "The *Challenger*, an Avoidable Tragedy: Largest Recovery Operation Ever Attempted: Salvage Experts Found Vital Clues," *Los Angeles Times*, June 10, 1986.

147 **nicknamed the *Newport Ship*:** "Newport Medieval Ship's Future Concerns Restoration Group," *BBC News*, August 8, 2014.

147 **sent the wreck to be scrapped:** Gilbert Kreijger, "*Costa Concordia* Dilemma: Salvage, Cut, or Sink?," Reuters, January 27, 2012.

148 **"Raise the *Titanic*?":** Paul Heyer, *Titanic Legacy: Disaster as Media Event and Myth* (New York: Praeger, 1995), 143.

148 **By January 1970, Doug Woolley began promising:** "*Titanic* May Make Port Here," *Liverpool Echo*, January 12, 1970.

148 **But the team had expanded:** "Age and Youth on *Titanic* Salvage," *Newcastle Evening Chronicle*, March 10, 1970.

148 **"So far, everything is proceeding absolutely on schedule":** "Now the Sad Super Liner Could Sail Again—Underwater in England," *Reading Evening Post*, April 14, 1970.

150 **"staked his existence on lugging":** "Could the *Titanic* Come to Liverpool?," *Liverpool Echo*, December 28, 1971.

151 **The USS *Monitor*, the famous ironclad:** "USS *Monitor*," Monitor National Marine Sanctuary, accessed November 16, 2020, https://monitor.noaa.gov/shipwrecks/uss_monitor.html.

152 **The bumps or depressions:** Gregory S. Mountain, "The Ocean Floor," Lamont-Doherty Earth Observatory, April 13, 2008, https://www.ldeo.columbia.edu/video/the-ocean-floor.

152 **discovered that undersea topography:** National Oceanography Centre, "Slope on the Ocean Surface Lowers the Sea Level in Europe," Phys.org, Janu-

ary 29, 2015, https://phys.org/news/2015-01-slope-ocean-surface-lowers-sea.html.

153 **after Malaysian Airlines flight 370 disappeared:** Ean Higgins, *The Hunt for MH370* (Sydney: Pan Macmillan Australia, 2019).

## CHAPTER 8: TAKE ALL THE BODIES AND TREAT THEM WITH RESPECT

156 **"The launching of a ship":** "Queen of Seas Dwarfs Men O' War," *The Leader-Post* (Toronto), September 27, 1938.

156 **Elizabeth grabbed a bottle of Australian wine:** John Honeywell, "Anniversary Memories of a Royal Liner That Couldn't Wait to Be Launched," Captain Greybeard, September 27, 2013, https://www.captaingreybeard.com/2013/09/anniversary-memories-of-a-ship.html.

156 **the band played:** "Her Majesty Launches *Queen Elizabeth* 1938," *British Pathé*, https://www.britishpathe.com/video/her-majesty-launches-queen-elizabeth.

157 **the ship caught fire in the harbor of Hong Kong:** "Fire Breaks Out on Former RMS *Queen Elizabeth*," last updated January 7, 2020, https://www.history.com/this-day-in-history/fire-breaks-out-on-queen-elizabeth.

158 **Woolley arrived in Hong Kong:** "*QE* Dress Rehearsal for Big *Titanic* Lift," *Liverpool Echo*, April 7, 1972.

158 **"They treated me better than anyone":** Doug Woolley, telephone interview by author, September 22, 2020.

158 **it may have been arson:** "Arson Suspected in *Queen Elizabeth* Fire," *Progress Bulletin* (Pomona, CA), January 10, 1972.

158 **An investigation revealed:** "Arson Sank Luxury Liner," *Springfield Leader and Press*, July 19, 1972.

159 **"They told Doug to go to Southampton":** Clive Amphlett and Douglas John Faulkner-Woolley, *Titanic: One Man's Dream: Douglas John Faulkner-Woolley: His Claims on Britain's Two Most Famous Liners (QE1 and Titanic): A Biography* (Ilford, UK: Seawise Publication, 1998).

161 **From the airport, he found his way into London:** Doug Woolley, telephone interview by author, May 14, 2020.

163 **he won the kind of attention:** "What's Going On," *Austin American-Statesman*, November 13, 1977.

163 **a two-masted Byzantine vessel:** "Serçe Limanı Shipwreck Excavation," Institute of Nautical Archaeology, accessed July 13, 2021, https://nauticalarch.org/projects/serce-limani-shipwreck-excavation.

164 **"The ship is its own memorial":** John P. Eaton and Charles A. Haas, *Titanic: Destination Disaster: The Legends and the Reality* (New York: W. W. Norton, 1996), 161.

164 **"Take the archeologists in Egypt":** Stephen Spignesi, "An Expanded Interview with Douglas Faulkner-Woolley," February 20, 2012, https://stephenspignesi.weebly.com/on-the-titanic.

164 **"We will take all the bodies":** Doug Woolley, in-person interview by author, London, March 7, 2020.

165 **The *Britannic* had sunk:** Simon Mills, *Exploring the Britannic: The Life, Last Voyage and Wreck of Titanic's Tragic Twin* (London: Bloomsbury Publishing, 2019).

166 **"Cousteau Strikes Gold":** "Cousteau Strikes Gold Below Sea," Associated Press, July 14, 1976.

167 **a hundred wrecks in the Ottawa River:** "Wreck Dates Back to 1870," *The Ottawa Citizen*, July 16, 1973.

167 **a Cycladic-era trading vessel:** "Ancient Shipwreck Discovered in Sea," *Standard-Examiner* (Ogden, UT), September 21, 1975.

167 **fishing trawler sank near Cape Cod:** "Divers Find Cape Trawler," *The Boston Globe*, February 20, 1978.

167 **happened upon the *San Juan*:** "Wreck Believed from 1565," *The Gazette* (Montreal, Canada), November 16, 1978.

167 **Operation Drake discovered:** "1699 Scottish Shipwreck Discovered," *The Post-Crescent* (Appleton, WI), April 28, 1979.

168 **"On the bottom of the ocean are mineral deposits":** "Mining the Wealth of the Ocean Deep," *The New York Times*, July 17, 1977.

168 **the quantities underwater:** Wil S. Hylton, "History's Largest Mining Operation Is About to Begin," *The Atlantic*, January/February 2020.

168 **Hjalmar Thiel went to a part of the Pacific Ocean:** Olive Heffernan, "Seabed Mining Is Coming—Bringing Mineral Riches and Fears of Epic Extinctions," *Nature*, July 24, 2019, updated August 16, 2019.

169 **Howard Shatto realized that if a captain:** Surender Kumar, *Dynamic Positioning for Engineers* (Boca Raton, FL: CRC Press, 2020).

169 **a strand of spaghetti:** Suzanne O'Connell, "Scientists Have Been Drilling into the Ocean Floor for 50 Years—Here's What They've Found So Far," *The Conversation*, September 26, 2018, https://theconversation.com/scientists-have-been-drilling-into-the-ocean-floor-for-50-years-heres-what-theyve-found-so-far-100309.

170 **a blown-out well off Santa Barbara, California:** Robert Olney Easton, *Black Tide: The Santa Barbara Oil Spill and Its Consequences* (New York: Delacorte Press, 1972).

170 **twelve wells off the Louisiana coast:** "Offshore Oil Blaze Blasted Out After Burning Since Last Dec. 1," *The New York Times*, April 13, 1971.

170 **a semisubmersible called the *Deepwater Horizon*:** Evan Thomas and Daniel Stone, "Black Water Rising," *Newsweek*, June 7, 2010.

172 **Massachusetts-based Woods Hole Oceanographic Institution:** "History & Legacy," Woods Hole Oceanographic Institution, accessed July 14, 2020, https://www.whoi.edu/who-we-are/about-us/history-legacy.

172 **"I looked up and saw this huge ball of fire":** Gerry Haden, "Palomares Bombs: Spain Waits for US to Finish Nuclear Clean-Up," *BBC News*, October 22, 2012.

173 **the support cables holding the *Alvin* snapped:** "Happy Birthday *Alvin*: 50 Years of Discovery," The Maritime Executive, June 4, 2014, https://www.maritime-executive.com/article/Happy-Birthday-Alvin-50-Years-of-Discovery-2014-06-04.

175 **held that the ship had sustained:** "Hunt for Tomb of the *Titanic*," *The Sydney Morning Herald*, November 10, 1979.

175 **any survey "had to go through me":** Doug Woolley, in-person interview by author, London, March 7, 2020.

175 **a model for forecasting the breakdown:** Keith Muckelroy, "The Integration of Historical and Archaeological Data Concerning an Historic Wreck Site: The 'Kennemerland,'" *World Archaeology* 7, no. 3 (1976): 280–90.

176 **"Only complete excavation can preserve":** Muckelroy, "The Integration of Historical and Archaeological Data Concerning an Historic Wreck Site: The 'Kennemerland.'"

## CHAPTER 9: PEOPLE THINK SINKING SHIPS IS EASY

177 **a crashed Aer Lingus plane:** *Historical Diving Times: The Newsletter of the Historical Diving Society* (Southampton, UK: Historical Diving Society, 2007).

178 **the story of the *General Grant* grew darker:** Madelene Ferguson Allen and Ken Scadden, *The* General Grant*'s Gold: Shipwreck and Greed in the Southern Ocean* (Auckland, New Zealand: Exisle Publishing Limited, 2009).

178 **a French ship called the *Anjou*:** Charles H. Lagerbom, "Siren Call of the *General Grant*: Shipwreck and Gold Fever with a Maine-Built Ship," *Coriolis: The Interdisciplinary Journal of Maritime Studies* 8, no. 2 (2018).

179 **he brought in Woolley:** Doug Woolley, telephone interview by author, September 22, 2020.

179 **"I got the sense":** Doug Woolley, telephone interview by author, August 16, 2020.

180 **He read dozens of books:** "Lord Mersey's Report on the Loss of the 'Titanic,'" *Nature* 89 (1912): 581–84, doi:10.1038/089581d0.

181 **Grattan mapped out his plan:** "Hunt for Tomb of the *Titanic*," *The Sydney Morning Herald*, November 10, 1979.

182 **large-holed nets with fine mesh:** Craig McClain, "An Empire Lacking Food," *American Scientist*, November–December 2010.

182 **"Most species are rare":** Robert R. Hessler and Peter A. Jumars, "Abyssal Community Analysis from Replicate Box Cores in the Central North Pa-

cific," *Deep Sea Research and Oceanographic Abstracts* 21, no. 3 (1974): 185–209, doi:10.1016/0011-7471(74)90058-8.

182 **to the seafloor under Antarctic ice:** "Unknown Animals Found Under Antarctic Ice," *Green Bay Press-Gazette*, November 3, 1977.

182 **While anchored near the Galápagos:** Sharon Proctor, "Waterworld," *The Province* (Vancouver, BC), March 7, 1982.

182 **it rivaled the rich biodiversity:** "Exploring the Deep Ocean Floor: Hot Springs and Strange Creatures," USGS, last updated June 24, 1999, https://pubs.usgs.gov/gip/dynamic/exploring.html.

183 **By illustrating the wonders:** Rachel Carson, *The Sea Around Us* (Oxford, UK: Oxford University Press, 1951).

184 **man-made reefs could help:** Frank W. Steimle and Richard B. Stone, *Bibliography on Artificial Reefs* (Washington, DC: Coastal Plains Center for Marine Development Services, 1973).

184 **Alabama Department of Conservation initiated:** Recreational Fisheries Management Subcommittee of the Technical Coordinating Committee, Gulf States Marine Fisheries Commission, "A Profile of Artificial Reef Development in the Gulf of Mexico," December 1993.

184 **Florida officials were sinking**: Donald W. Pybas, *Atlas of Artificial Reefs in Florida* (Gainesville: Florida Sea Grant College Program, University of Florida, 1991).

185 **thousands of tires:** "Florida Retrieving 700,000 Tires After Failed Bid to Create Artificial Reef," *The Guardian*, May 22, 2015.

185 **"People think sinking ships is easy":** Tim Mullane, telephone interview by author, October 16, 2020.

187 **The *Queen Mary* was launched:** John Maxtone-Graham, *Queen Mary 2: The Greatest Ocean Liner of Our Time* (New York and Boston: Bulfinch, 2004).

188 **it's been falling apart:** Julia Hatmaker, "How 'America's Flagship' Ended Up Decaying Outside of an Ikea: The Story of the SS *United States*," *Penn Live—Patriot-News*, November 22, 2016.

188 **"one of the world's most dangerous jobs":** Peter Gwin, "The Ship-Breakers," *National Geographic*, April 2014.

190 **John Grattan's proposed survey:** Christopher Dobson, "How We Will Find the Tomb of the *Titanic*," *Now!*, October 19–25, 1979.

190 **Disney and National Geographic, joined forces:** John P. Eaton and Charles A. Haas, *Titanic: Destination Disaster: The Legends and the Reality* (New York and London: W. W. Norton, 1996).

191 **the tall ship *Bounty*:** National Transportation Safety Board, "Marine Accident Brief: Sinking of Tall Ship *Bounty*," February 6, 2014, https://www.ntsb.gov/investigations/AccidentReports/Reports/MAB1403.pdf.

193 **a Florida-based company:** Dan Spurr, "Unsinkable: The History of Boston Whaler," Professional Boatbuilder, March 26, 2019, https://www.proboat.com/2019/03/unsinkable-the-history-of-boston-whaler.

193 **Point Nemo is the unofficial dumping ground:** Dave Mosher, "A Spacecraft Graveyard Exists in the Middle of the Ocean—Here's What's Down There," *Business Insider*, October 22, 2017, https://www.businessinsider.com/spacecraft-cemetery-point-nemo-google-maps-2017-10.

194 **"Time is working against us":** Douglas Woolley, letter to Philip Slade, Ilford, UK, 1979, Woolley personal archives.

## CHAPTER 10: A HEIFER CORRALLED IN A BOX CANYON

195 **"Well, you won't get very far":** Doug Woolley, telephone interview by author, April 16, 2020.

195 **That was how he had built:** Jack Grimm and William Hoffman, *Beyond Reach: The Search for the Titanic* (New York and Toronto: Beaufort Books, 1982).

196 **He had begun at age eleven:** Robert Mcg. Thomas Jr., "Jack F. Grimm, 72, Is Dead; A Seeker of Oil and Legends," *The New York Times*, January 9, 1998.

196 **they spent their honeymoon:** Edward Helmore, "Obituary: Jack Grimm," *Independent* (UK), January 19, 1998, https://www.independent.co.uk/news/obituaries/obituary-jack-grimm-1139665.html.

196 **"This is the ark":** Thomas, "Jack F. Grimm, 72, Is Dead."

196-197 **cursory study of weather patterns:** Grimm and Hoffman, *Beyond Reach*.

197 **"I've never heard a mouth":** Doug Woolley, in-person interview by author, London, March 6, 2020.

198 **long-disputed CQD mystery:** Grimm and Hoffman, *Beyond Reach*.

200 **The *City of Cairo* was sunk:** Joel Gunter, "Record Dive Rescues $50m Wartime Silver from Ocean Floor," *BBC News*, April 15, 2015, https://www.bbc.com/news/world-africa-32316599.

200 **When the *Rhakotis* was off the coast:** Guðmundur Helgason, "*City of Cairo*: British Steam Passenger Ship," uboat.net, accessed November 14, 2020, https://uboat.net/allies/merchants/ship/2383.html.

202 **"When you see something that could be a shipwreck":** Gwilym Ashworth and John Kingsford, telephone interview by author, April 4, 2020.

202 **"Then fire the scientists":** Paul Heyer, *Titanic Century: Media, Myth, and the Making of a Cultural Icon* (Santa Barbara, CA: Praeger, 2012), 169.

202 **the search ship stopped for a night:** Kathleen Maxa, "The Texas Tycoon in Search of the *Titanic*," *The Washington Post*, June 21, 1981.

203 **"I think we got that heifer":** "'We Think We've Got the *Titanic*,'" *The San Bernardino County Sun*, August 16, 1980.

203 **"highly optimistic":** "*Titanic* Search Goes on in Spite of Some Damage," *The New York Times*, August 14, 1980.

203 **the crew had been surprised:** John Noble Wilford, "Scientists in Expedition Doubt *Titanic* Discovery," *The New York Times*, August 22, 1980.

204 **"The big money is to be made":** "What Really Happened to the Proudest Ship That Ever Sailed?," *The New York Daily News*, April 16, 1980.

205 **The club had sponsored ambitious expeditions:** "Historical Highlights," The Explorers Club, accessed March 12, 2021, https://www.explorers.org/about/history-including-famous-firsts.

205 **"The Explorers Club has always been associated":** "Roll Call for Adventure," *The Sydney Morning Herald*, November 1, 1980.

206 **a postal worker delivered a wad of letters:** Maxa, "The Texas Tycoon in Search of the *Titanic*."

207 **Thomas Edison once experimented:** Walter von Baeckmann, Werner Prinz, and Wilhelm Schwenk, eds., *Handbook of Cathodic Corrosion Protection* (Oxford, UK: Elsevier Science, 1997), 12.

207 **The system became known as cathodic protection:** M. Heldtberg, I. D. MacLeod, and V. L. Richards, "Corrosion and Cathodic Protection of Iron in Seawater: A Case Study of the *James Matthews* (1841)," *Proceedings of Metal* (2004): 75–87.

208 **has made the Chuuk Lagoon:** Dan E. Bailey, *World War II Wrecks of the Truk Lagoon* (Redding, CA: North Valley Diver Publications, 2001).

208 **"The metal wears down over time":** Chris Selman and Allison Selman, telephone interview by author, April 29, 2020.

209 **the National Oceanic and Atmospheric Administration keeps a list:** "Potentially Polluting Wrecks in U.S. Waters," National Marine Sanctuaries, last updated July 31, 2017, https://sanctuaries.noaa.gov/protect/ppw/welcome.html.

209 **were linked to the *Luckenbach*:** "SS *Jacob Luckenbach* Oil Removal," *Incident News*, May 23, 2002, https://incidentnews.noaa.gov/incident/1008.

209 **eight hundred twenty-five tons of coal:** Greg Ward, *The Rough Guide to the Titanic* (London: Rough Guides, 2012).

210 **"The second stage of our expedition":** Harihar Krishnan, "Millionaire Jack Grimm Said Sunday He Is Confident," UPI, April 12, 1981.

210 **Bobby Blanco, a Cuban-born gambler:** Grimm and Hoffman, *Beyond Reach*.

211 **enticed thousands of people:** Grimm and Hoffman, *Beyond Reach*.

211 **the problem of credit:** Grimm and Hoffman, *Beyond Reach*.

212 **"Wouldn't it be great":** Grimm and Hoffman, *Beyond Reach*.

213 **answering questions like a celebrity:** Grimm and Hoffman, *Beyond Reach*.

214 **The uneven terrain underscored:** Melvin Porter, "A Geological Study of the *Titanic* Shipwreck Site," *Owlcation*, February 18, 2017, https://owlcation.com/stem/The-Geology-of-The-Titanic-Shipwreck-Site.

215 **the overwhelming sensation:** Grimm and Hoffman, *Beyond Reach*.

217 **the *Ocean Ranger*:** Keith Collier, "The Loss of the *Ocean Ranger,* 15 February 1982," Heritage Newfoundland & Labrador, 2010, updated October 2016, https://www.heritage.nf.ca/articles/economy/ocean-ranger.php.

217 **the *Mekhanik Tarasov,* sank too:** *Summary Report of Investigation into the Circumstances Attending the Foundering of the Russian Vessel Mekhanik Tarasov in the North Atlantic, on February 16, 1982* (Canada: Canadian Coast Guard, n.d.).

218 **A debris-tracking program:** Laura Parker, "Ocean Trash: 5.25 Trillion Pieces and Counting, but Big Questions Remain," *National Geographic*, January 10, 2015, https://www.nationalgeographic.com/science/article/141211-ocean-plastics-garbage-patches-5-gyres-pollution-environment.

219 **Three studies in 2015:** Brian Howard Clark, "Five Trillion Pieces of Ocean Trash Found, but Fewer Particles Than Expected," *National Geographic*, December 13, 2014, https://www.nationalgeographic.com/science/article/141211-ocean-plastics-garbage-patches-5-gyres-pollution-environment.

219 **a group of South African researchers:** Peter G. Ryan et al., "Rapid Increase in Asian Bottles in the South Atlantic Ocean Indicates Major Debris Inputs from Ships," *Proceedings of the National Academy of Sciences of the United States of America* 116, no. 42 (2019): 20892–97, doi:10.1073/pnas.1909816116.

219 **"It's inescapable that it's from":** Aylin Woodward, "New Research Suggests We Might Be Thinking About the Ocean Plastic Problem All Wrong—Trash Dumped from Ships Could be a Major Culprit," *Business Insider*, October 4, 2019, https://www.businessinsider.com/plastic-pollution-ocean-comes-from-ships-trash-2019-10.

219 **Newer regulations:** *United States Code, 2006 Edition, Supplement 4, Title 33—NAVIGATION AND NAVIGABLE WATERS 2010* (Washington, DC: Government Printing Office, 2006).

220 **projections that ocean freight:** "Global Freight Demand to Triple by 2050," The Maritime Executive, May 27, 2019, https://www.maritime-executive.com/article/global-freight-demand-to-triple-by-2050.

220 **called the *Ever Given*:** "*Ever Given*: Ship That Blocked Suez Canal Sets Sail After Deal Signed," *BBC News*, July 7, 2021.

221 **the *Maersk Essen*, a cargo ship:** Ann Koh, "*Maersk* Loses 750 Containers as Seas Hammer LA-Bound Ship," *Bloomberg*, January 22, 2021, https://www.bloomberg.com/news/articles/2021-01-22/maersk-loses-750-containers-in-pacific-ocean-on-bad-weather.

222 **"Wait a minute":** Grimm and Hoffman, *Beyond Reach*, 180.

## CHAPTER 11: ALL THESE MOTHS DRAWN TO THE SAME FLAME

223 **a triple-screw system:** Wyn Craig Wade, *The Titanic: Disaster of the Century* (New York: Skyhorse Publishing, 2012).

223 **designed a propeller like a screw:** Don Leggett, *Shaping the Royal Navy: Technology, Authority and Naval Architecture, c.1830–1906* (Manchester, UK: Manchester University Press, 2016).

224 **a tug-of-war:** Leggett, *Shaping the Royal Navy.*

225 **fueling endless debates:** Mark Chirnside, "The Mystery of *Titanic*'s Central Propeller," *Encyclopedia Titanica*, accessed February 11, 2021, https://www.encyclopedia-titanica.org/mystery-titanic-central-propeller.html.

225 **"Play it again":** Jack Grimm and William Hoffman, *Beyond Reach: The Search for the Titanic* (New York and Toronto: Beaufort Books, 1982), 180.

225 **Grimm called for Captain Armand:** Grimm and Hoffman, *Beyond Reach.*

228 **Mearns found the *Lucona*:** David L. Mearns, *The Shipwreck Hunter: A Lifetime of Extraordinary Discovery and Adventure in the Deep Seas* (New York: Pegasus, 2018).

229 **"You'd be surprised":** David Mearns, telephone interview by author, June 19, 2020.

231 **ordered Grimm and Mearns:** "*Titanic* Salvaging Still Held Up in Court," *Daily Press* (Newport News, VA), October 1, 1992.

231 **"You have all these moths":** David Mearns, telephone interview by author, June 19, 2020.

232 **"Third time's the charm":** "Texan Set for Third Hunt," *The Vancouver Sun*, July 12, 1983.

233 **an eighteen-inch-long miniature replica:** "Taking a Leaf from a Gilded Book," *The Philadelphia Inquirer*, December 15, 1983.

233 **"We are all very pleased":** "*Titanic* Expedition Crew Still Hopeful," *News Pilot* (San Pedro, CA), August 4, 1983.

235 **several loyalists rowed him:** John Toohey, *Captain Bligh's Portable Nightmare: From the* Bounty *to Safety—4,162 Miles Across the Pacific in a Rowing Boat* (New York: Skyhorse Publishing, 2019).

235 **thirty million people each year:** "Number of Ocean Cruise Passengers Worldwide from 2009 to 2020," *Statista*, August 4, 2021, https://www.statista.com/statistics/385445/number-of-passengers-of-the-cruise-industry-worldwide.

236 **"Realistically it can't be done":** David Mearns, telephone interview by author, June 19, 2020.

237 **"I don't believe in the curses":** William Broad, "Effort to Raise Part of *Titanic* Falters as Sea Keeps History," *The New York Times*, August 31, 1996.

237 **the Luxor hotel in Las Vegas:** "*Titanic*'s Largest Recovered Artifact 'The Big Piece' at *Titanic*: The Artifact Exhibition," *Vegas News*, September 9, 2011, https://vegasnews.com/56809/titanics-largest-recovered-artifact-the-big-piece-at-titanic-the-artifact-exhibition.html.

238 **In the final report:** Neil Monney and Parks Stephenson, "A Technical Plausible Scenario for the Salvage of the *Titanic*," U.S. Naval Academy, 1977.

239 **the ship was lowered and raised:** *Raise the Titanic*, directed by Jerry Jameson (ITC Entertainment, Associated Film Distribution, 1980).

## CHAPTER 12: MAN IS NEVER LOST AT SEA

246 **Two days earlier:** William Broad, "Wreckage of Titanic Reported Discovered 12,000 Feet Down," *The New York Times*, September 3, 1985.

247 **One of the crafts:** Robert D. Ballard and Rick Archbold, *The Discovery of the Titanic* (New York: Warner Books, 1987).

247 **"We went smack-dab":** "U.S.-French Team Finds Wreckage of *Titanic*," *The Washington Post*, September 3, 1985.

247 **"We realized we were dancing":** "*Titanic*: The Untold Story," *CBS News*, December 9, 2018, https://www.cbsnews.com/news/titanic-the-untold-story.

248 **Ballard had made a bargain:** John Roach, "*Titanic* Was Found During Secret Cold War Navy Mission," *National Geographic*, November 21, 2017, https://www.nationalgeographic.com/history/article/titanic-nuclear-submarine-scorpion-thresher-ballard#close.

249 **Ballard occasionally referred to Michel:** Ian Coutts and Robert D. Ballard, *Titanic: The Last Great Images* (Philadelphia: Running Press, 2008).

250 **"This allows us to open":** Lily Rothman, "See Photos of the Wreck of the *Titanic* When It Was First Discovered," *Time*, September 1, 2015.

251 **"In a way I am sad":** Robert Ballard, "A Long Last Look at *Titanic*," *National Geographic*, December 1986.

251 **"It's sort of like I just married someone":** Steven Biel, *Down with the Old Canoe: A Cultural History of the Titanic Disaster*, updated edition (London: W. W. Norton, 2012), 203–4.

252 **Ballard confirmed plans to sell:** William Broad, "Finder of *Titanic* Aims to Capitalize," *The New York Times*, September 11, 1985.

252 **visiting the wreck to collect artifacts:** Paul Gilbert, "Explorer Defends Himself Against *Titanic* Criticism," *The Miami Herald*, April 9, 1987.

252 **"You don't go to the Louvre":** Hampton Sides, "*Titanic*: The Most Famous Wreck," *National Geographic*, October 29, 2015.

253 **"What do I have to do":** Ryan D'Agostino, "The Man Who Found the *Titanic* Is Not Done Yet," *Popular Mechanics*, August 4, 2015, https://www.popularmechanics.com/adventure/a16715/bob-ballard-oceanographer-titanic.

254 **"It's all in there":** Doug Woolley, in-person interview by author, London, March 6, 2020.

256 **"did my work for me":** Clive Amphlett and Douglas John Faulkner-Woolley, *Titanic: One Man's Dream: Douglas John Faulkner-Woolley: His Claims on Britain's Two Most Famous Liners (QE1 and Titanic): A Biography* (Ilford, UK: Seawise Publication, 1998).

257 **"I've sacrificed friends":** Doug Woolley, in-person interview by author, London, March 7, 2020.

259 **"a man is never lost at sea":** Ernest Hemingway, *The Hemingway Collection* (London: Scribner, 2014).

## CHAPTER 13: A REDDISH STAIN IN THE MUD

261 **Harland & Wolff continued building ships:** "Robert Ballard: Restore the *Titanic*," *National Geographic Society* Resource Library, accessed February 8, 2021, https://www.nationalgeographic.org/media/ng-live-titanic-video-gallery.

262 **"It's been a hundred years":** "Robert Ballard: Restore the *Titanic*."

262 **Eric Kuhne to design:** "*Titanic* Belfast by Eric R. Kuhne & Associates," AAS Architecture, December 1, 2013, https://aasarchitecture.com/2013/12/titanic-belfast-by-eric-r-kuhne-associates.html.

262 **Belfast officials thought:** "*Titanic* Belfast Had 800,000 Visitors in a Year," *BBC News*, April 23, 2013, https://www.bbc.com/news/uk-northern-ireland-22263955.

263 **a record long held:** Alex Lemaire, "The World's Most Valuable Shipwreck: The *Nuestra Señora de Atocha*," The Maritime Executive, July 20, 2020, https://www.maritime-executive.com/editorials/the-world-s-most-valuable-shipwreck-the-nuestra-senora-de-atocha.

263 **"Anybody can make a plate or teacup":** Ozgur Ar, telephone interview by author, November 8, 2020.

264 **there are different *Titanic*s:** N'dea Yancey-Bragg and Corey Arwood, "The First Manned Dive to the *Titanic* in 14 Years Found a Wreck in 'Shocking' Decay," *USA Today*, August 21, 2019.

266 **the grand staircase, no longer exists:** Don Lynch and Ken Marschall, *Ghosts of the Abyss: A Journey into the Heart of the Titanic* (Boston: Da Capo Press, 2003).

267 **Asisi offered an artistic interpretation:** "Yadegar Asisi's *Titanic* 360 Panorama in Leipzig," *UR Design*, April 7, 2017, https://www.urdesignmag.com/art/2017/04/07/titanic-360-panorama-leipzig.

268 **the 2010 Woods Hole visit:** Steve Szkotak, "New *Titanic* Expedition Will Create 3D Map of Wreck," Associated Press, July 27, 2010.

268 ***Halomonas titanicae*:** "New Species of Rust-Eating Bacteria Destroying the *Titanic*," *Live Science*, December 6, 2010, https://www.livescience.com/9079-species-rust-eating-bacteria-destroying-titanic.html.

269 **In 2001, a New York couple:** "*Titanic* Set for Shipwreck Wedding," *Wired*, July 16, 2001.

269 **a company called Ocean Gate Expeditions:** Ben Finley, "As the *Titanic* Decays, Expedition Will Monitor Deterioration," Associated Press, June 30, 2021.

269 **"Visitors do more damage":** William Broad, "Scientists Warn That Visitors Are Loving the *Titanic* to Death," *The New York Times*, August 9, 2003.

269 **he was apparently "amazed":** "*Titanic* Deteriorates," *Herald & Review* (Decatur, IL), September 4, 2003.

270 **"This is terrible":** "Return to *Titanic* Raises Sensitive Issues for Salvage Company," *The Journal News* (White Plains, NY), August 7, 2000.

270 **still chugging along in 2010:** Richard Foot, "Race to Map *Titanic* Before It Dissolves," *The Calgary Herald*, August 21, 2010.

270 **In 2019, she gave it:** Gina Martinez, "Divers Visited the *Titanic*'s Wreck for the First Time in Over a Decade," *Time*, August 22, 2019, https://time.com/5658903/titanic-wreck-deteriorating.

270 **the tannins in leather:** Colin Pearson, *Conservation of Marine Archaeological Objects* (London: Butterworth-Heinemann, 1998), 128.

271 **"It'll be there for hundreds of years":** David Mearns, telephone interview by author, June 19, 2020.

272 **"at best an embarrassment":** Clive Palmer, TV interview by Lisa Wilkinson, *Today Queensland*, February 4, 2016.

273 **"When the day comes, and it most surely will":** *60 Minutes Australia*, "*Titanic* Clive," July 30, 2013.

273 **Sarel Gous had dreamed of reconstructing:** Henry McDonald, "*Titanic* Survivor Condemns Plan to Rebuild Ship as Cashing in on Tragedy," *The Guardian*, December 30, 2000.

274 **the box-office haul in China:** Laura Burkitt, "Why China Loves 'Titanic' So Much: A Theory," *The Wall Street Journal*, April 25, 2012.

276 **"The only reason I'm still alive":** Doug Woolley, in-person interview by author, London, March 7, 2020.

# Index

Note: Italicized page numbers indicate material in photographs or illustrations.

## About the Author

**Daniel Stone** is a former staff writer for *National Geographic* and White House correspondent for *Newsweek*. He holds degrees from UC Davis and Johns Hopkins University, where he now teaches environmental science and history. He lives in Santa Barbara with his wife and two sons, one of whom is a dog.